Cost–benefit analysis

A practical guide

Cost–benefit analysis

A practical guide

Second edition

Michael Snell

Published by Thomas Telford Limited, 40 Marsh Wall, London E14 9TP.
www.thomastelford.com

Distributors for Thomas Telford books are
USA: Publishers Storage and Shipping Corp., 46 Development Road, Fitchburg, MA 01420
Australia: DA Books and Journals, 648 Whitehorse Road, Mitcham 3132, Victoria

First published 1997

Also available from Thomas Telford Limited
Planning Major Projects. R.J. Allport. ISBN 978-0-7277-4110-3
Asset Management. C. Lloyd (ed.). ISBN 978-0-7277-3653-6
Art and Practice of Managing Projects. A. Hamilton. ISBN 978-0-7277-3456-3

www.icevirtuallibrary.com

A catalogue record for this book is available from the British Library

ISBN: 978-0-7277-4134-9

Typeset by Academic + Technical, Bristol
Index created by Indexing Specialists (UK) Ltd, Hove, East Sussex
Printed and bound in Great Britain by CPI Antony Rowe, Chippenham and Eastbourne

Contents

Preface to the first edition   ix

Preface to the second edition   xi

Introduction   xiii

1   Basic principles of cost–benefit analysis   1
    1.1   A set of tools for guiding decisions   1
    1.2   A common unit of measurement   2
    1.3   Types of cost–benefit analysis   3
    1.4   The decision-making context   6
    1.5   Value judgements   7
    1.6   Analysts, decision-makers and transparency   7
    1.7   Risk, uncertainty and pragmatism   8
    1.8   Stages of the planning–design–implementation process   8
    1.9   The main stages of a cost–benefit analysis   10
          Notes   11

2   Quantifying and valuing the costs and benefits   13
    2.1   Costs and benefits in the with-project and
          without-project situations   13
    2.2   Costs and benefits that are sometimes not counted   16
    2.3   Defining the common unit of measurement   24
    2.4   Shadow pricing: practical application   26
    2.5   Rationale of shadow pricing   28
    2.6   Valuation of particular kinds of costs and benefits   31
    2.7   Double-counting   36
          Notes   38

3   Resources and time: discounting and indicators   40
    3.1   Time preference   40
    3.2   Discounting   42
    3.3   Indicators and their calculation   46

3.4	Length of analysis period	52
3.5	Inflation, depreciation and loans	54
3.6	Choice of indicator for particular purposes	57
3.7	Sensitivity tests	61
3.8	Terminology and reporting	66
3.9	The appropriate discount rate	68
	Notes	75
4	**Application to particular fields**	**78**
4.1	General	78
4.2	Agriculture	82
4.3	Health projects and programmes	90
4.4	Water supply, sanitation, public health	94
4.5	Education	96
4.6	Flood and coastal protection, environment, recreation	98
4.7	Transport	111
4.8	Energy	115
4.9	Urban and regional planning	117
	Notes	117
5	**Reporting and presentation**	**120**
5.1	Definition of the project	122
5.2	Methodology and assumptions	123
5.3	Statement of price basis	123
5.4	Probabilistic aspects of benefit estimates	124
5.5	Treatment of sensitive demand forecasts	125
5.6	Optimisation of timing	125
5.7	Financial analysis	127
5.8	Size optimisation and choice of indicator	128
5.9	Graphical presentation of sensitivity test results	129
5.10	Sensitivity tests in text	131
5.11	Summary	133
	Note	133
6	**Checklist**	**134**
	APPENDICES	**139**
A	**Uncertainty, probability and risk**	**141**
A.1	Basic concepts	141
A.2	Attitudes to uncertainty	149
A.3	Ways of handling uncertainty in CBA	153

	A.4	Summary and recommendations	159
	A.5	Further reading and information sources	159
		Notes	160
B	**The domestic pricing and foreign exchange numéraires**		**163**
	B.1	Introduction	163
	B.2	Traded and non-traded goods	164
	B.3	The domestic pricing system	164
	B.4	The foreign exchange system	165
	B.5	Comparison of the two systems	166
	B.6	Choice of system	167
	B.7	Further reading and information sources	169
		Notes	170
C	**Ways of estimating economic prices**		**172**
	C.1	Introduction and underlying principles	172
	C.2	Market prices with adjustments	175
	C.3	Opportunity cost approaches	178
	C.4	Stated preferences; contingent valuation	182
	C.5	Proxy markets and other revealed-preference approaches	184
	C.6	Shadow pricing as an instrument of policy	188
	C.7	Choice of valuation method	188
	C.8	Benefit transfer	189
	C.9	Further reading and information sources	190
		Notes	191
D	**Distribution of costs and benefits between people**		**195**
	D.1	General distribution adjustments	195
	D.2	Income weighting	196
	D.3	Use of distributional weighting	198
		Notes	198
E	**Choice of discount rate**		**200**
	E.1	Introduction	200
	E.2	Individual and social time preference	201
	E.3	Opportunity cost of capital	202
	E.4	Capital rationing device	203
	E.5	Risk and the discount rate	204
	E.6	Synthesis	205
	E.7	Further reading and information sources	207
		Notes	209

F   Multicriterion decision analysis   **212**
　　F.1   Introduction   212
　　F.2   Basic concepts and techniques   214
　　F.3   Non-numerical methods   220
　　F.4   A simple numerical method   221
　　F.5   Other methods   225
　　F.6   Uncertainty and incomplete information   229
　　F.7   Further reading and information sources   229
　　　　　Notes   230

G   The effects method   **232**
　　　　　Notes   233

H   Model answers to readers' worked examples   **235**

I   Worked example   **241**
　　I.1   Introduction   241
　　I.2   The project   241
　　I.3   Economic prices   243
　　I.4   Farm level financial analysis   245
　　I.5   Economic benefits   246
　　I.6   Economic costs   249
　　I.7   Cost–benefit indicators   250
　　I.8   Sensitivity tests   251
　　I.9   Conclusion   253
　　I.10   Postscript   254

J   Discounting tables   **256**

K   Bibliography   **267**

　　Glossary/index   **281**

Preface to the first edition

This book aims to meet the needs of engineers, planners, environmentalists and other professionals who occasionally use cost–benefit analysis (CBA) or related project appraisal techniques, or who prepare inputs for such analyses. It will also be useful to people who take decisions that are partly guided by CBA – officials, politicians and representatives of particular groups affected by a project or a plan. The book will help those who lead multidisciplinary teams or edit multifaceted documents in which CBA is one aspect among many. It can also serve as a source book for economists needing to deepen or update their knowledge of CBA.

The book is intended to be used in two complementary ways:

- firstly, as a learning tool whereby the reader can, starting with basic arithmetic but no training in economics, understand the rationale of CBA and learn to carry out many of the operations
- secondly, as a reference book, which a planner, engineer, economist or other professional, who has already learned the techniques and is applying them, can turn to for a reminder of basic principles or for details on a particular sort of analysis that he has not done before.

Readers will not become economists by studying this book, but they will understand more of what economists are saying. They will also be able to discuss CBA with economists and ask them the right questions, which is often necessary because economists cannot be expected to understand all the technical and planning implications of the numbers and concepts they manipulate. Some practising economists learn a great deal about other disciplines, and communication will be furthered if engineers and others can go some way to meet them. After working through this book, readers should be able to undertake some kinds of CBA unaided, especially when the more complex parameters such as shadow price factors and discount rates have already been fixed by others. The book includes detailed coverage of uncertainty and risk, and of multicriterion analysis.

This book originates from my own felt need. While practising as a civil engineer and becoming increasingly involved in feasibility studies, I found I was checking, re-working and editing CBAs originally prepared by economists. When the economist was absent or innumerate, I had to compute the results, and often I had to interpret them for clients or decision-makers who could not understand the terminology. I was always conscious of the partial nature of my understanding of what I was doing. Eventually I went to university and asked the academics to fill the gap. I found that there was a shortage of books, and of teachers, able to explain CBA to people who are fully numerate and understand the details and contexts of planning and project decisions, and who also have some experience of finance, but who lack a formal training in economics. Since taking that postgraduate course I have been able, in some situations, to function as an economist, and often to influence and supervise CBAs done by others.

In 1990 I began to give courses in CBA to graduate engineers and other professionals as part of the in-house training programme of the consulting firm I work for, Mott MacDonald. I learned what sorts of explanation work, and what sorts bring a glazed expression to the audience's eyes. The firm's formal feedback system enabled the participants to comment on my teaching and to contribute to its improvement. I am grateful to Mott MacDonald for permission to take advantage of this experience in writing this book, and for permission to use here some of the material I developed for the course.

I would like to thank many who have helped me to an understanding of CBA and project economics. In particular I am very grateful to the economists and other professionals who have commented on parts of the draft: Colin Bruce, Jim Perry, Clare Rhodes-James, Maureen Sibbons, Brigitte Snell and Michael Yaffey, and especially to the publisher's reviewer, David Potts of the Development and Project Planning Centre, University of Bradford, and to my colleague, mentor and constructive critic Chris Finney. I have not always followed their advice, and indeed this would not be possible because they disagree with each other. The publisher's referees provided useful ideas on content and structure. Finally, I thank the 330 victims of my Mott MacDonald graduate course on project economics held over the last eight years, for their comments and questions, and for their invaluable, and sometimes unintended, feedback on how not to explain things. None of these people, however, should be blamed for the book's shortcomings.

Preface to the second edition

The principles of cost–benefit analysis (CBA) have not changed in the years since the 1997 edition was published, but a few useful books and guidelines have been published by others, and these are referred to in this new edition. References to internet sources have also been added.

Some passages have been clarified or extended, and in particular the appendix on multicriterion analysis has been reworked in the light of my experience in the intervening years.

CBA is now more often carried out using spreadsheet or workbook software than it was in 1997, and the guidance has been updated with reference to current commonly used software, including a warning about a particular pitfall in that software.

I acknowledge with gratitude the encouragement of many readers and users of the first edition, especially Chris Finney, Jim Winpenny and Mike Chauhan. They have confirmed that CBA can be understood and used by people who are not formally trained in economics, and that practising economists also appreciate a clear explanation of the subject and its many applications.

Introduction

Cost–benefit analysis (CBA) is a decision-guiding tool that has traditionally been regarded as a branch of economics. This book has been written primarily for non-economists, by a non-economist, because I have found that such people can and should use the tool. To work well in guiding decisions, CBA needs to be understood by the decision-makers and other parties affected by a decision, and this book aims to spread such understanding.

The chapters constitute a complete account of CBA as it affects people who practise it or use its results, but avoiding most of the theoretical details. The basics are given in Chapters 1 to 3. Chapter 4 introduces the special aspects of a number of sectors and project types, using these also to introduce some analytical devices that can apply to many fields. Because the result of a CBA is only useful when a decision-maker understands it and accepts its validity, Chapter 5 gives some guidance on the presentation of a CBA in a report or similar document, or in a personal presentation. Finally, Chapter 6 provides a checklist of steps to be taken and questions to be dealt with, both in the conducting and the reporting of a CBA.

The chapters are designed to be read sequentially, each providing the concepts needed to understand the ones that follow, but they can also be used for reference. Many concepts are explained or illustrated in boxes or figures within the chapters, so as not to interrupt the flow of the main text. Numerical examples are included in the explanatory sections, which the reader using the book as a learning tool can work through if desired, and some chapters end with more extensive questions for which model answers are presented in an appendix. Appendix I contains a worked example of a complete CBA report.

The more specialised theory and the finer distinctions are mostly in the appendices and endnotes, with selective references to the literature. For those using the book as a reference work, the Glossary/Index gives pointers to both chapters and appendices; it brings together cross-references to the text and definitions of the main terms and abbreviations. Some of the terms defined in the Glossary/index are identified in the text by being printed in italics.

Notes are collected together at the end of each chapter or appendix. They are mainly references to published literature, and readers should generally ignore them unless they need those references. Many sections and appendices end with pointers to further reading and useful references.

Most of the illustrations and examples are from my own experience, some being composites of more than one actual analysis. Because of this, many of them concern water resources development of some kind, but the concepts and methods are applicable to many fields, and throughout the book the emphasis is on basics and wide application. Because of confidentiality constraints, only those examples and case studies that are already in the public domain are named.

To avoid the use of 'he or she', the reader, analyst or decision-maker is sometimes referred to as female and sometimes as male; any bias in the distribution of usage is unintended.

1

Basic principles of cost–benefit analysis

... whether the pleasure of making a daisy-chain would be worth the trouble of getting up and picking the daisies.[1]

1.1   A set of tools for guiding decisions

We all make decisions every day, whether on matters of life and death or on more mundane questions like whether or not to take an umbrella when setting out for a walk. Consciously or not, before deciding on a course of action we often ask ourselves 'Is it worth it?' When we do that, we are using *cost–benefit analysis*. Having formulated a possible course of action (a project), we assess the *cost* it will incur and the *benefits* it will bring, balance the one against the other, perhaps consider other influences, and then ask: 'Is the benefit worth the cost?' So cost–benefit analysis (CBA) is, above all, a set of tools for guiding decisions.

In decisions like the umbrella one, the balancing of costs and benefits is done in a moment and often unconsciously, while in professional contexts the use of CBA is highly organised, quantitative and carefully argued. Yet many features are common to both scenarios. Like any other tool, CBA can be well used or badly used: some critics of CBA quote examples of its misuse as criticisms of the tool itself, which is as clever as saying chisels are bad because some people use them as screwdrivers. The use of CBA often involves *uncertainty* – the consequences of setting out on the walk without the umbrella usually cannot be firmly predicted, and a probability estimate is needed for that decision just as it is for the design of a spillway or an earthquake-threatened structure.

To use CBA quantitatively for the guidance of making a decision, there are a few basic steps in every case:

(a) define the decision, normally between a possible course of action and its alternative
(b) fix, at least tentatively, the decision criteria

(c) estimate the cost of taking that course of action
(d) estimate the benefits it would bring
(e) weigh up the costs and benefits by means of some quantitative indicator
(f) consider uncertainty and the range of possible outcomes
(g) apply the criteria, consider the CBA alongside other relevant decision guides, such as environmental aspects, and make the decision.

CBA can be used to guide many sorts of decision. Typically, it is used either for a yes/no decision on whether a single course of action will be undertaken or not, or to choose between two or more competing courses of action. In each case the decision can be a major one, such as whether or not to construct a tunnel under the English Channel, or what medical interventions to use for a certain condition, or a minor one, such as whether to equip a certain bridge with painted or galvanised handrails.

1.2   A common unit of measurement

In order to ask 'Is the benefit worth the cost' in any quantitative way, we need to have the cost and the benefit expressed in some sort of *common unit of measurement*. It is not much help to decision-makers to say a bridge will cost £8 million and its benefit will be to save 15 minutes per working day for 3500 people; the benefits have to be put in the same unit as the costs, however approximate the methods involved, before this bridge can be compared with other bridges or with new schools and hospitals competing for the same resources.

As costs are most easily and habitually measured in terms of money, it is usual, although not in principle essential, to use money or something similar as the common unit of measurement. Some welfare economists speak of a unit of utility called a 'utile', but this is practically never used for CBA. So the numerical part of a CBA is nominally conducted in dollars, euros, pounds, rupees, pesos, etc. In many kinds of CBA, as will be described in this book, the unit is not really money, because some kind of economic or social pricing has been used rather than ordinary financial or market prices, but the unit is still called by the name of a currency. This can cause confusion and only careful labelling and explanation can define the unit properly; for example, 'Euros at 2015 prices'. The non-financial kinds of pricing are the subject of much of this book; they usually aim to give expression to people's collective preferences, using money merely as an instrument for that purpose.

Much of the work of conducting a CBA lies in putting the costs and the benefits into the common unit. When the various costs and benefits are of different kinds, such as bridge construction costs and journey-time reductions, the processes needed are called *valuation methods*, and these are discussed in Chapter 2. When they occur at different times, the relative value of resources at different moments has to be taken into account by *discounting*, which is the subject of Chapter 3. Many analyses require both these procedures.

1.3   Types of cost–benefit analysis

The main types of decision guided by CBA are:

- *investment-type* or *yes/no decisions*, such as whether or not a single project or course of action will be undertaken
- *design-type* or *either/or decisions*, such as which of several possible projects should be implemented, or the choice between two or more alternative ways of achieving some technical goal (competing courses of action).

A special kind of design-type decision is the case of *technically mutually exclusive* courses of action. In this case the alternatives are such that only one can be chosen, for technical reasons rather than shortage of resources. Examples are the choice between a higher and a lower dam at the same site, between a two-lane and a three-lane road, between materials (one cheap, one durable, one beautiful), or between methods (one labour-intensive, one machinery-intensive, or one faster than another). In general, we have to quantify both costs and benefits, but in some analyses, called *least-cost* or minimum-cost ones, all the competing courses of action produce the same benefits. Common examples are alternative conveyors for water transfer (long canal, short tunnel; small pipeline with high pumping costs, large pipeline with low pumping costs; etc.) and thermal versus hydroelectric power. Being common to all courses, the benefits then do not need to be evaluated at all, nor even quantified beyond ensuring that they are really the same for all alternatives. This kind of analysis is often easier to do than a CBA, as benefits are usually more difficult to value than costs. It often helps to formulate a decision in this form, to simplify the analysis. (If there is more than one kind of benefit we can use so-called weighted cost-effectiveness analysis, which in this book is treated in the context of multicriterion analysis in Appendix F. When alternatives are almost identical but not quite, for instance the product of one alternative is the

3

Table 1.1 Types of analysis for technically mutually exclusive courses of action

Type of analysis	Treatment of costs and benefits
Cost–benefit analysis (CBA)	Both vary between courses of action; both are valued in a common unit
Cost-effectiveness analysis (CEA)	Both vary between courses of action; both are quantified, but not in a common unit
Least-cost analysis (special case of CBA and CEA)	Costs vary and are valued; benefits are identical for all courses of action and do not need to be valued

same quantity as the product of the other alternative but in a slightly different quality, small compensating adjustments can be made, rather than abandoning the least-cost type of analysis.)[2]

A least-cost analysis is also a special case of another type of analysis, *cost-effectiveness analysis*. In the more general case of a cost-effectiveness analysis, the benefits and costs are not expressed in a common unit of measurement, but both are quantified and then different courses of action are compared on the basis of the relative costs per unit of benefits. For a health programme the effectiveness measure might be the number of children inoculated. For a water supply scheme we might calculate the cost per cubic metre of water supplied, or per household supplied. For flood protection the ratio might be the cost per person protected from a specific sort of flood, and for any sort of land improvement it might be the cost per hectare. Cost-effectiveness analysis is often used in fields where the benefits are difficult to value economically, such as healthcare or education.[3] Table 1.1 summarises these kinds of analysis for technically mutually exclusive courses of action.

Analyses are also distinguished according to the identity of the group of people on whose behalf they are carried out, or whose interests are to be taken into account in the decision. The main categories are financial, economic and social CBA.

1.3.1 Financial CBA

A financial analysis concerns the financial position of a person, firm or organisation, so that both costs and benefits are measured in terms of money spent or received by that party, regardless of whether the prices are a good reflection of true value. This kind of analysis includes taxes and subsidies and is not concerned with price distortions. (Financial analysis of projects, in a wider sense than CBA, is described in Yaffey (1992) and Potts (2002).)

1.3.2 Economic CBA

An economic analysis concerns the welfare of a defined group of people, usually a nation. Although market prices and money flows are usually the starting point for the quantification of costs and benefits, they are considered to be an imperfect representation of the group's best interests, and are therefore adjusted in various ways, such as so-called shadow pricing, which is described in Chapter 2. The adjustments are often towards efficiency prices, corresponding to the concept of a perfect market that achieves the best possible allocation of resources by the interaction of supply and demand. (In economic theory, prices are the signals that mediate the optimisation of allocation, and efficiency prices are those that would obtain in a perfect market.) For better or worse, real market prices are distorted by political actions, taxes and subsidies, monopolies, imperfect competition, incomplete or misleading information, fashion, habit and many other factors. In CBA, the purpose of economic pricing, sometimes called *efficiency pricing*, is to adjust the market or financial prices in order to correct for these distortions and arrive at the prices that a perfect market would arrive at. As the concept of a perfect market is not a purely technical one, this usually involves value judgements, and can only be partially and approximately achieved; the way it is done is one of the main sources of subjectivity, and hence dispute, in economic CBA. Occasionally, for some kinds of costs or benefits, economic prices are estimated directly rather than by adjusting financial prices.

1.3.3 Social CBA

This is a rare but distinct form of CBA. In a social CBA the analyst goes further and adjusts the prices by which costs and benefits are valued, so as to reflect priorities and policies that no market would reflect, not even a perfect market. There is no implication that such prices are anything but subjective applications of value judgements, usually of an explicitly political nature. Examples are adjustments to give advantage to certain regions of a country, or to certain population groups, such as the rural poor, while the adjustment most often discussed in textbooks refers only to income (see Appendix D).

In practice, the distinction between economic (efficiency) CBA and social CBA is often blurred, both kinds being referred to as economic analysis.[4] In this book, reference to economic analysis includes social CBA unless an explicit distinction is made. In many contexts the word *social* only means that collective interests are being emphasised rather than individual ones. The term social analysis, as distinct from social CBA, means something quite different, usually of a sociological nature.

The terms 'economic analysis' and 'economic appraisal' are sometimes used more widely than here, to include not only the economic CBA of a project but also the financial CBA for various parties involved in it. The rationale for this is that the project will fail unless participants have something to gain by using it; this theme is discussed further in later chapters. In this book the term 'economic CBA' or 'economic analysis' has the meaning defined above, although a project assessment may involve more than just an economic CBA.

1.4   The decision-making context

Although CBA is a tool for guiding decisions, it is seldom the only guide to a particular decision. Most decisions are made on several criteria, some of which may in practice be unacknowledged and never explicitly presented (perhaps because decision-makers cannot defend them). Examples of other criteria which, rightly or wrongly, are often used alongside CBA are:

- regional policy (a political desire to develop some regions)
- social policy (a political desire to favour some groups of people, such as disadvantaged minorities)
- environmental criteria
- self-interest (a decision-maker's desire to improve his own wealth or position)
- habit
- risk aversion.

It is generally desirable that decision-making be clearly explained and justified, even if the reasons are confidential. Sometimes, when the matter is a very simple one, the reasons and criteria underpinning a decision merely need to be explained in words. In more complex cases, some decision analysis framework is desirable, both for arriving at the decision and for subsequently explaining and defending it. Appendix F describes some multicriterion decision analysis methods suitable for the systematic and transparent use of CBA alongside other criteria.

In principle, many criteria that do not immediately seem to be of a financial or economic nature can nevertheless be brought into a CBA by means of special valuation methods. For instance, it is technically possible to reflect regional policy or preferential treatment of a particular minority in the valuation of costs or benefits (social pricing). Environmental criteria can, in principle, also be applied inside CBA. The fact that this is possible does not necessarily make it desirable. It may be better to apply such criteria explicitly alongside the CBA rather than to apply them within it, because such separate treatment is more transparent and open to discussion.

1.5   Value judgements

Except in the simplest of cases, CBA involves *value judgements*. In the rare case of social CBA these judgements have to be explicit, but in economic analysis some people try to give the impression that the analysis is value-free and can only be done in one way with a single outcome. In fact the price adjustment process usually involves some subjective factors, and it is unusual for two independent economists to produce identical sets of economic prices. Academic economists have, from time to time, attempted to construct theoretical frameworks that claim absolute validity, independent of value judgements. The necessary axioms and assumptions, however, inevitably incorporate ethical principles and value judgements, and so any economic or social CBA incorporates such judgements and is in some degree subjective. Practitioners and users of CBA should neither try to hide this nor apologise for it, but should make the criteria and value judgements explicit: such an analysis is called 'transparent'.[5]

1.6   Analysts, decision-makers and transparency

The need for transparency in decision guides arises from the roles of *decision-maker* and *analyst*. Sometimes these roles are combined in one person or party, but usually there is one person or group of people responsible for making (and later defending and/or implementing) the decision, and another person or group responsible for analysing the relevant factors and advising the decision-maker. This latter party is here called the 'analyst'.

There is often a danger that the analyst will usurp part of the decision-maker's role, by hiding the value-laden criteria and procedures of the analysis and presenting only the result, as if it were value-free. This is unprofessional or dishonest, or both. A good decision guide is transparent in that the criteria and methods are explicitly declared; the analyst explains clearly all the assumptions made and the rationale behind each one. If the recommended decision does not feel right to a decision-maker or critic, she can then examine those criteria and methods and question them individually, rather than vaguely expressing misgivings about the final result. The analyst can facilitate this by including sensitivity analyses (described in later chapters of this book) in his analysis. A decision based on a transparent analysis is likely to be owned by the decision-maker, and such decisions are more likely to be effectively implemented than are decisions based on expert advice that the decision-maker does not understand, and therefore cannot check or verify.[6]

1.7   Risk, uncertainty and pragmatism

An analysis is only useful as a decision guide if it is realistic. Technical solutions and project components must be not only technically feasible but also socially and politically feasible. Estimates should be realistic ones in the real world, not wishful expressions of what might be a good thing if only it could be achieved. The first requirement is for a practical approach on the part of the analyst, backed up with experience and judgement, rather than textbook theory. The second is a careful and systematic approach to *risk* and *uncertainty*. These are the subject of Appendix A and are also mentioned in many sections of other chapters. If an estimate is very approximate, the analyst should say so and explain how it was arrived at and why things might turn out differently, analysing the consequences by means of *sensitivity tests*, as explained in this book. If some aspect of a project is important but difficult to forecast or quantify, it is much better to include it in the analysis, in a candidly approximate way, than to leave it out. Only an analysis of uncertainty can show up the value of flexibility and adaptability to varying conditions, which are important characteristics of a project.

1.8   Stages of the planning–design–implementation process

A typical engineering project or undertaking goes through a succession of stages, from initial concept, through design and implementation, to operation, and beyond, such as:

(a) Project identification: the project is merely a concept, such as a transport link across an estuary, the nature of which (tunnel, bridge or ferry) is still undecided.

(b) Pre-feasibility study: many options, few data; serves to decide broadly what sort of crossing and whether it is worth pursuing further.

(c) Feasibility study and project appraisal: fewer fundamental options, more field data, more detail; serves to define the project more closely and guides the decision of whether to implement it or not.

(d) Final design and production of tender documents.

(e) Tender evaluation, leading to signing of a contract (at stages (d) and/or (e) the CBA of the feasibility study may be reworked with better cost estimates, even with revised benefit estimates, and the decision to proceed may be reviewed).

(f) Detailed design and working drawings.

(g) Construction and commissioning.

(h) Operation and use of the project.

This is usually not the end of the story, as over the years the project suffers

(i) decay

(j) changing needs

which may eventually necessitate its

(k) rehabilitation (requiring another round of studies, decisions, design, contracts, commissioning, etc.).

Unless maintenance is very good, and needs remain unchanged, the decay and rehabilitation cycle can be repeated several times, and eventually the project may be closed down, leaving either some assets with significant residual value, or some dangers and future costs, such as waste disposal or the decommissioning of an old nuclear power station.

CBA is useful at several of these stages.[7] At stages (b) and (c), the pre-feasibility and feasibility studies, CBA is used to compare the merits of basic design alternatives at various levels of detail (bridge versus tunnel, road tunnel versus rail tunnel, steel versus concrete tunnel lining, etc.), and also to indicate whether or not the project is worth pursuing further. The consequences of the latter decision are far weightier at stage (c) than at stage (b), as in the pre-feasibility study there is only the cost of a feasibility study to be lost by a wrong decision, but at the feasibility study stage the consequences of a mistaken decision to proceed are much more significant, which is why a feasibility study is usually more thorough and more expensive than a pre-feasibility study.

9

At each stage the CBA reflects the current levels of detail and of precision.

At the design stages, (d) and (f), CBA is used for some of the lower-level design decisions, such as pipeline size or the choice of materials; these technical decisions are often of the least-cost type. CBA is seldom used in tender evaluation, but it can be.[8]

If and when a project needs rehabilitation, most of the previous decision stages occur again, with the corresponding use of CBAs in studies, design and evaluation. In this case the definition of the project often depends crucially on what is assumed to happen in its absence, called the *without-project situation*.

1.9   The main stages of a cost–benefit analysis

To provide a framework for the discussion of techniques and details in the remaining chapters of this book, the main activities of a CBA are listed below. This list amplifies the basic concepts introduced in Section 1.1; it refers to an economic analysis, and the details for a financial one would differ in a few respects.[9] A least-cost economic analysis would omit the quantification of benefits. Various terms are mentioned without definition here, but they are all defined in the subsequent chapters and also in the Glossary/Index at the back of the book.

(a) Define the decisions that are to be guided. When the analysis concerns a project, the project must be precisely defined, which can usually best be done by defining its purpose and then the *with-project* and *without-project situations*.

(b) Define the group of people whose point of view is to be applied.

(c) Decide the criteria and parameters, such as:
   - project life, or period to be used for analysis
   - discount rate (see Chapter 3)
   - categories of benefits and costs (some things can be treated either as positive benefits or as negative costs)
   - adjustments – shadow pricing, omission of transfer payments, etc. (see Chapter 2).

(d) Calculate the economic benefits attributable solely to the decision or the project for each year of its life, i.e. the *incremental benefit* (the benefits of the with-project situation minus any benefits that would arise in the without-project situation).

(e) Similarly, calculate economic costs:
   - initial (capital) costs
   - recurrent (annual) costs

- replacement costs, e.g. replace pumps in years 16, 31, etc.; in each case the *incremental cost* is with-project cost minus the without-project cost, for each year of the project life.

(f) Formulate the net benefit stream (optional):
- list the benefits and costs year by year for the whole analysis period
- calculate the *net benefit* (benefit minus cost) for each year (the list of net benefits is sometimes called *cash flow*, although in an economic analysis it is not cash but resources valued at economic prices).

(g) Carry out the arithmetical economic analysis, i.e. *discount* the benefit and cost (or net benefit) streams and calculate indicators, as explained in Chapter 3.

(h) Carry out *sensitivity tests* (showing what would happen to the indicators if the parameters and assumptions were different from base-case values).

(i) Assess other benefits or costs that have been excluded from the CBA.

(j) Report the whole analysis:
- define the project
- declare the purpose of the analysis (what decision, whose point of view)
- declare and explain the assumptions and criteria
- give the results, base case and then sensitivity tests
- comment on them, draw conclusions from the sensitivity tests, but avoid prejudging the decision
- give enough information for the report reader to check the arithmetic.

Some guidelines on this reporting and presentation are in Chapter 5, and Chapter 6 gives a checklist based on this list of stages. Chapters 2 and 3 present the theoretical and practical basis for valuation and for discounting, respectively, and Chapter 4 discusses applications to various fields.

Notes

1 *Alice's Adventures in Wonderland*, Lewis Carroll (C. L. Dodgson), Macmillan, 1865, Ch. 1.
2 Belli *et al.* (2001, Ch. 7) (full references are given in Appendix K).
3 A discussion of three kinds of cost-effectiveness analysis is given in van Pelt (1994, Section 3.6).

4 The distinction between economic and social analysis is explained in Irvin (1978, Section 4.08) and in van Pelt (1993, Section 3.2.3); the latter places social CBA in the context of multicriterion analysis.

5 The most important value judgement is probably the assumption that willingness-to-pay is the right basis for guiding decisions, despite the obvious fact that it varies with different people's wealth, i.e. their ability to pay (see Appendices C and D). The moral foundations, and issues concerning the comparison of one person's welfare with another's, are discussed in academic books such as those by Adler and Posner (2006), Brent (2006) and Zerbe and Bellas (2006).

6 Transparency concerning assumptions is emphasised in the World Bank handbook (Belli *et al.*, 2001, p. 16).

7 The World Bank handbook places heavy emphasis on this, mentioning decisions on a project's location, scale, timing, sequencing, choice of technology, choice of beneficiaries and types of outputs (Belli *et al.*, 2001, start of Ch. 3).

8 Use of discounting in evaluating construction bids is discussed and advocated by Hardy *et al.* (1981).

9 Some analysts begin by analysing the financial flows for the agency implementing a project, and then base the economic analysis on this; the World Bank handbook recommends such a practice, and emphasises the financial effects on various parties alongside the economic analysis (Belli *et al.*, 2001).

2

Quantifying and valuing the costs and benefits

2.1 Costs and benefits in the with-project and without-project situations

Among the early steps in a numerical CBA is the task of identifying the costs and benefits associated with the project or decision, in preparation for putting a value on them. Examples of the sorts of things ('goods and services' in economic terminology) that may need to be quantified, and questions that may need to be answered, are:

- For an engineering construction, how many tons of cement and of steel will be needed, how many hours of skilled and of unskilled labour, how many hours of work by what machines?
- How much land will a project use up (take away from other uses), and what sort of land?
- How much energy and labour will the operation of the project need?
- How many journey hours and how many accidents will be saved by a road improvement?
- How many patients will benefit from a medical policy change, and how will their lives be changed?
- How many children will be affected by an education project, and with what effect on their lives and their families?
- How many extra tons of wheat will be grown as a result of the reha-bilitation of an irrigation scheme?
- How much energy will a power station produce, and when? (What time of year, what time of day?)
- To whom does each benefit accrue, and who bears each cost?

Some of these costs and benefits are directly related to the project's purpose and are obvious enough, but others are more or less unavoid-able side-effects. Some costs and benefits have market prices, but some do not. Typical adverse unpriced effects are noise and pollution,

Cost–benefit analysis – A practical guide
ISBN: 978-0-7277-4134-9

while positive ones can include improvement of the visual merit of an area by a fine bridge, or provision of recreation facilities by a dam.

Some adverse side-effects can be mitigated by special measures, and these measures should be built into the project before it is analysed, so that the cost of the mitigating measures are counted and the reduction of adverse effects is also taken into account.

The key to correct and complete identification of costs and benefits is the clear definition of each project (or, more generally, each course of action). This is normally done by defining the *with-project situation* and the *without-project situation*. These are projections into the future, with all the usual difficulties and uncertainties of estimating and forecasting. The without-project situation, sometimes but not always equivalent to the do-nothing scenario, is not usually the same as a continuation of the present situation, although it may be similar to it. Some examples are:

(a) For the economic analysis of a project to rehabilitate an old and run-down irrigation scheme, the definition of the without-project situation is crucial because the project's benefits are primarily the difference between the two situations' agricultural outputs (this can be a small difference between large numbers, and consequently liable to large uncertainties in estimation). One definition of the without-project situation is that the scheme will continue to deteriorate and its output will decline from year to year into the future. Another possible assumption would be that, even in the absence of the proposed rehabilitation project, some maintenance and repair would be carried out so as to keep the scheme from further deterioration and maintain the present level of output. Under the first of these definitions there would be little or no without-project costs, but the without-project benefits would decline over the years; under the second there would be significant ongoing without-project costs but the without-project benefits would be maintained. Either definition will serve, but it would be wrong to mix them by assuming constant without-project benefits without allowing for the necessary costs.[1]

(b) In the analysis (financial or economic) of a new road bridge or a road improvement project, the definition of the without-project situation will largely determine how the analysis is done. Will the traffic continue to use the old road network as it stands, even if traffic flows increase? Or should the without-project situation be defined as some alternative set of actions to alleviate the road problems? Will the traffic volumes be the same in the with-project and without-project situations?

(c) The objectives of a certain coastal protection scheme included preventing the breaching of a line of dunes behind which are an inhabited area and an environmentally valuable habitat, as well as reinstating an eroded beach and protecting a road along the dunes. If the without-project situation had been defined as the continuing and increasing risk of the dunes and road being breached, the loss of the hinterland would have had to be valued. Instead, by defining the without-project situation as one in which the road was always maintained and repaired so as to prevent the breaching of the dunes, the valuation of hinterland effects was made unnecessary, although the without-project situation included significant ongoing costs on the road.

In each case the chosen definition of the without-project situation should be a likely, or at least credible, alternative course of action, i.e. one that might really happen if the project does not go ahead. Once the two situations have been clearly defined, the analyst must carefully list all the costs and benefits of both. Within limits, it is reasonable (as in example (c)) to adjust the definitions so as to make the quantification of costs and benefits easier; if the two situations can be defined so as to make some particular impact identical in the two, then that impact will not feature in the incremental costs and benefits which alone determine the result, and will not need to be quantified and valued.

Part of the work of defining the with-project and without-project situations is to consider whether a project's outputs add to total production of the relevant goods, or substitute for (displace) goods that would have been produced or provided anyway. If the latter is true, the effects on the other providers of the goods may need to be considered too. If the project is small relative to the relevant sector in the relevant country or region, such effects will usually be negligible, but if the project is large in such contexts it will often be necessary to analyse a wider unit, such as a whole sector. For example, to help decide about the implementation (both yes/no and timing) of a very large multistage hydropower project (here named Sila) in a relatively undeveloped country, we analysed the development of the generating capacity of the whole national grid over the next half-century. After many plausible development plans had been analysed, involving Sila and/or several other hydro and thermal plants, the yes/no question was addressed by comparing the best with-Sila plan against the best

15

without-Sila plan, and the optimum Sila start date was shown by comparing the different with-Sila plans.[2]

As well as being very clearly defined, each project or course of action should (to a reasonable degree of approximation) be internally optimised and, if it has discrete or separable components, optimally packaged. These concepts are explained in Section 3.6 in the context of indicators.[3]

2.2   Costs and benefits that are sometimes not counted

2.2.1   Guiding principles

There are some sorts of costs and benefits that are counted in one kind of analysis and not another, and some that are never counted but which can cause confusion. In principle, the inclusion or exclusion of anything can be decided by going back to the principles discussed in Chapter 1, and asking two basic questions:

- What decision is this CBA intended to guide?
- Which defined entity is this CBA being done for or on behalf of? (Normally a commercial entity or financially distinct agency in a financial analysis, and a nation in an economic analysis.)

This section applies the principles to some common categories that can cause confusion.

2.2.2   Transfer payments

These are financial payments between different parties within the defined entity on whose behalf the economic analysis is done, for instance a nation in the case of an economic analysis. From the point of view of the nation as a whole, such payments have no significance. Individuals and subgroups may experience them as costs and benefits, but for the nation collectively (the national economy) they are like funds transferred from one accounting category to another within one firm, or cash moved from my left pocket to my right pocket. So transfer payments, whether linked with costs or with benefits, must be omitted from any CBA.

A particular class of payment that is sometimes a transfer payment, and sometimes not, is *taxes* (including customs and excise duties), together with *subsidies*, which are in effect like negative taxes. In a financial CBA for an entity that pays taxes or duties, or receives subsidies, where the common unit of measurement is money in that entity's possession, they are not transfer payments: they are ingoings and outgoings like any others, and should be included in the financial CBA. In an economic analysis, however, where the defined entity is a nation, taxes and subsidies are transfer payments and should not be counted in a CBA. The government is only a subgroup within the nation, and these payments between the government and particular citizens do not increase or decrease the welfare of the nation as a whole. (They may, however, serve to guide shadow prices, see Section 2.5.)[4]

2.2.3 Non-market and unpriced costs and benefits

These are consequences of a decision, such as a project's inputs or outputs, that have no market and do not give rise to financial payments to or from the relevant entity; they are sometimes called *externalities* (Box 2.1), and another common, but vague, term is *intangibles*. Whether they are counted in a CBA depends, again, on what sort of CBA it is.

As emphasised above, a financial analysis is carried out for an entity (individual, firm or institution) on the sole criterion of its money, so the common unit is money in the possession of that entity. Costs borne by others should not be counted, unless those others are financially compensated by the entity, nor should benefits accruing to others be

---

Box 2.1 Externalities

The usual economist's definition of an externality is an effect or product that does not have a market or money price, so that people or entities that make decisions on the basis of financial analyses and criteria will not take it into account. In a CBA context, however, externalities are sometimes defined as the costs and benefits that will arise from the execution of a project under consideration but which do not impinge on parties directly involved in the project itself (the practical effect of this depends on what *directly* means).[5]

In an economic or social CBA an effect that has no money valuation but does impinge on the society concerned would be an externality under the first definition but not under the second, if 'the parties directly involved' is interpreted broadly. Because of this ambiguity the term 'externality' is usually avoided in this book.

---

counted unless they pay the entity for those benefits. All the central entity's financial outgoings and takings consequent on the project or course of action must be counted, regardless of whether they are a good or fair valuation of something, and no others, irrespective of the justice or equity of the matter. If a factory pollutes a river and raises costs or spoils facilities for downstream users, and gets away with it without any payment, the downstream effects do not count in the factory's or its promoter's financial CBA; to these parties the pollution is an externality. If, however, the government imposes some sort of pollution levy, pollution licence fee or fine, or if the factory owner pays financial compensation to downstream sufferers, or if he builds them a sports club to ease his conscience and reduce opposition, or if he installs water treatment facilities to reduce the pollution, those costs and payments are counted because the pollution effects have now been internalised into the entity's unit of measurement; they now have financial prices to the factory owner, and are no longer externalities.

An economic analysis, however, is one done on behalf of a defined social group, usually a nation. So any cost borne by some member of that group should be counted, whether or not any money changes hands between that person and the project or its promoter. Similarly, all benefits should be counted, even if no one pays anything for them. All significant effects should be included in the analysis (unless they occur outside the region or country and do not affect the members of the defined group at all). This may be a counsel of perfection; in practice, some effects may have to be left outside the CBA as unquantifiables, in which case they should be described to the decision-maker alongside the CBA results, whether in a formal multicriterion analysis (see Appendix F) or just in verbal reporting. Unquantifiable effects are not the same as externalities, although some effects fall in both categories and the terms are sometimes loosely used as synonyms. The ways in which values can be put on non-market goods is discussed in Section 2.6.

A significant category of unpriced costs is work or goods provided voluntarily or otherwise unpaid. This includes the work of volunteers, and of family members working on farms. It can include equipment given to a school or hospital by a charity, the local community or any other donor. Such inputs should be costed (generally by the opportunity cost approach described in Sections 2.5 and 2.6) and included in an economic analysis, although they do not appear in a financial analysis.

2.2.4 Sunk costs

Sunk costs are those that are already incurred or irreversibly committed to before the moment of the decision which the CBA is to guide, so that they cannot be affected by that decision. The CBA is trying to weigh up the advantages and disadvantages of the decision's going one way or the other, and the sunk costs therefore have nothing to do with it. This simple argument ought to convince all doubters, but calls for the inclusion of sunk costs in a CBA are still heard from time to time (Box 2.2). The distinction is especially important in analysing rehabilitation projects,

---

Box 2.2 An attempt to include sunk costs in a CBA

An Asian government started building an irrigation scheme with its own resources, but ran out of funds. It approached an international development bank for a loan, and the bank commissioned a consultant to quantify the costs and benefits of completing the scheme and to conduct an economic CBA. The consultant rightly ignored the cost of the work already done and included only the costs of completion, but included all the benefits of the operation of the project, because the unfinished works were incapable of producing any benefits. The analysis indicated that the completion of the scheme was economically justified. Some members of the development bank's appraisal team asked for an analysis with the sunk costs included, to find out whether the scheme would have been a worthwhile one if it had been analysed before any construction began. The consultant said that was another analysis, not related to the decision he had been retained to guide. Eventually an estimate was made of the economics of the whole construction, including the sunk costs, but this was clearly separated from the CBA that guided the real decision. The result of the second analysis was unfavourable, but it was still worthwhile, with the unfinished works already standing, to finish them and enjoy the benefits. The scheme as a complete unit was inherently uneconomic, but its completion, which was what had to be decided about, was economic.

Postscript: Another financier offered funds and the development bank pulled out, but the funds did not materialise and the government attended to other things. The scheme remained unfinished and the farmers continued growing rain-fed crops between the half-built canals.

---

where the key is the correct definition of the without-project situation. If the physical assets representing sunk costs could in fact be reused in another place or for another purpose (e.g. railway rails that could be lifted), they do have some opportunity cost and may need to be valued after all; in effect such costs are not irretrievably sunk.[6]

2.2.5   Best-estimates and contingencies

In a CBA all the costs and benefits should be valued according to the analyst's best-estimates of the relevant quantities or payments, excluding any extra allowances that might be made for budgeting purposes. It is important not to confuse three distinct things that are all sometimes called contingencies:

(a) *Physical contingencies* are extra allowances, beyond the estimator's best-estimate, to make an underestimate less likely than an over-estimate (the concept is explained in terms of probabilities in Box 2.3).

Box 2.3 Best-estimates and physical contingencies

When an engineer estimates the quantity of work needed for a project, for example the volume of earth-moving for an embank-ment, there is always a degree of uncertainty. The engineer may say the estimate is $2500\,\mathrm{m}^3$, but in fact, whether it is stated or not, a probability distribution is implied, something like that shown in Figure 2.1. The average or expected value of the final quantity is $2500\,\mathrm{m}^3$, but there is a finite possibility that it will be as low as $2300\,\mathrm{m}^3$ or as high as $2800\,\mathrm{m}^3$. In principle it is possible to quantify the uncertainty and say, for example, that there is a 5% chance that the volume will be less than $2350\,\mathrm{m}^3$, and a 5% chance that it will be over $2730\,\mathrm{m}^3$. If one is preparing a budget, one is painfully aware that an underestimate will be far more embar-rassing and troublesome than an overestimate, and so one will normally quote a figure on the high side of the probability distribu-tion, say $2700\,\mathrm{m}^3$, even though one's expectation or best-estimate is that it will be only $2500\,\mathrm{m}^3$. The extra allowance, here $200\,\mathrm{m}^3$, is often called a *contingency allowance* or *physical contingency*, even if it is expressed only in money terms and arises purely from possible changes in unit rates without any change in physical quantities.

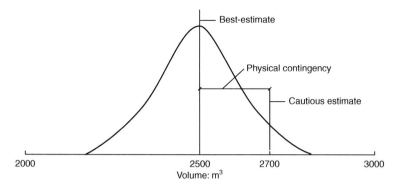

Figure 2.1 Best-estimates and physical contingencies

(b) Sums included in a preliminary cost estimate to cover minor construction or cost items that will certainly occur but have not yet been detailed, such as roads or fences in the preliminary estimate for a power station. These are not the subject of uncertainty in the sense of those in Box 2.3, and should preferably not be called contingencies at all (although they often are).

(c) So-called *price contingencies* are an accounting device to include the effect of inflation on budgets. These are explained in Box 3.4 in Chapter 3, page 56.

The treatment of these sorts of allowance follows from the purpose of the analysis:

(a) For a budget, the inclusion of a physical contingency allowance is respectable and sensible, provided that the estimate is accompanied by a clear statement of what sort of estimate it is. However, for a CBA physical contingencies should not be included, because cost estimates are going to be compared with the best-estimates of other parameters.

(b) An estimate of the cost of *miscellaneous items* that have not yet been detailed, but which will definitely arise, is not a contingency allowance in this sense, but a part of the best-estimate; it should be included in a CBA, however approximately, along with any other real costs. In pre-feasibility studies, where data are scarce and cost estimating is fairly crude, miscellaneous items are often costed by adding a rule-of-thumb percentage to the total estimate of the main items; the fact that this is arithmetically similar to the addition of a percentage physical contingency should not be allowed to make people think it is the same thing. Despite these

21

arguments for keeping them separate, physical contingencies and miscellaneous items are often mixed or combined. This is discussed further in Appendix A.[7]

(c) Price contingencies concern inflation and involve the adding together of numbers in different units. As explained more fully in Chapter 3 they have no place in a CBA, either financial or economic or social, which should be conducted in constant-value terms that cut out inflation altogether.

2.2.6 *Depreciation*

Depreciation is a financial device. It is a sum of money shown in the accounts of a firm or other entity to allow for the fact that the assets (property, equipment, etc.) are subject to wear and tear. The book value of the assets is reduced from year to year as their physical condition declines, and they eventually have to be replaced or refurbished if the production of benefits is to continue. There are various different ways of computing depreciation allowances, and the results sometimes affect taxation. A firm may also earmark some of its reserves for future replacements.

Depreciation and associated accounting procedures are not costs or benefits for a CBA, even a financial one, because they are merely book-keeping entries.[8] If they cause changes in the entity's monetary ingoings or outgoings, for instance through tax rules, these changes will count in a financial CBA, but the changes are not themselves depreciation. A CBA of any kind should, however, take into account real replacement and refurbishment costs, as and when they occur.

2.2.7 *Loans, repayments and interest*

Often a firm, a government department or any other entity borrows money to finance a project. Starting at some agreed moment, the firm pays interest on the outstanding loan, and it also pays the originally loaned money back to the lender according to an agreed timetable. Sometimes a financial CBA will count all these payments as and when they occur, because such an analysis refers strictly to the monetary ingoings and outgoings of the entity. Alternatively, a financial analysis can be carried out before financing (i.e. as if the implementing agency were doing it without a loan), as a means of finding out what loan is required.

For an economic CBA, these monetary payments are generally not counted. Instead, the costs and benefits, after suitable valuation, are counted as and when they physically occur. (If the lender is outside the nation, one could argue that the loans and repayments should be counted in the economic analysis, from the nation's point of view, but this is not normally done.) An example is given, in the context of the timing of costs and benefits, in Section 3.5, and Chapter 5 gives an example of the graphical presentation of a financial analysis.

2.2.8 *Summary*

When there remains some doubt about whether any cost or benefit should be included in a CBA, it can usually be settled by going back to the definition of the entity on whose behalf the analysis is being carried out, and drawing a conceptual boundary around that entity or group of people. Any real benefit or cost affecting them, such as the food they eat, amenities they enjoy, cement they use up, labour they perform, or travel time they save, is counted. So is anything that they import from another place or export to another place. But money that some of them pay to others within the group is a transfer payment and does not count. (If the payment is related to some output of the project, the output needs to be valued, but the payment itself is not a benefit or cost of the project.)

Another difficulty may be that of determining what does or does not pertain to the project, and here the key is the clear definition of the decision to be guided by the CBA (every CBA is only a tool for guiding decisions), and within that the clear definition of the project. As in the discussion of sunk costs, the principle is that all costs and benefits that flow from that decision must be counted, and no others. So, if the definition of the decision and the project involves a without-project situation that itself has costs, the analysis must use the incremental cost, i.e. the difference between with-project costs and without-project costs, because that is what would follow from the decision.

For things that can be identified as transfer payments, non-market elements, sunk costs, contingencies, depreciation allowances, or loans and their repayments and interest, the above arguments can be summarised in a set of rules-of-thumb governing whether they should be in or out of a CBA. Such a rule set is given in Table 2.1.

Although these arguments determine what should and should not be counted in a fully detailed CBA of any type, a short-cut is sometimes possible through the use of the *least-bound approach*. If one is comparing the absolute merits of two or more projects or courses of action, one must use the central or best-estimate value of each parameter. But if one is

Table 2.1 Guidelines for the inclusion of categories in a CBA

Type of analysis	Financial	Economic
Transfer payment, e.g. taxes	In	Out
Non-market costs or benefits (sometimes called 'externalities')	Out	In (even if only approximately valued)
Sunk costs	Always out	
Best-estimates, including miscellaneous items	Always in (differentials between with-project and without-project situations)	
Physical contingencies (as defined above)	Always out[9]	
Price contingencies	Always out	
Depreciation allowances	Always out	
Loans and their interest and repayment	In, as and when paid[10]	Out, but include replacement costs

merely demonstrating that a project meets some set of minimum criteria, it may be sufficient to use a *lower-bound* approach. In effect one says:

> *I do not know (or do not have time to calculate) the most likely value of the benefits, but they must be at least this much.*

If, having done that, one finds the project justified, one can argue that the true benefits, known to be more than the figure used, do not need to be estimated any more precisely. When some categories of benefits are difficult to put a value on, this line of reasoning can greatly simplify an analysis, without making it any less convincing as a decision guide.

2.3   Defining the common unit of measurement

As explained in Chapter 1, the definition of the *common unit of measurement* (called the *numéraire*) depends first on the defined entity on whose behalf the analysis is done:

- in the case of a financial analysis, money in the possession of the entity
- in the case of an economic (or social) analysis, the welfare or interests of the defined entity or group, normally a nation.

The financial numéraire is straightforward enough, but in the second case there is a need for more precise definition, and there are two distinct ways of conducting the analysis.

24

The simplest economic numéraire is the *domestic pricing* one, which uses ordinary domestic prices as the starting point.[11] The common unit of measurement in this sort of analysis is, in effect, the money held by the citizens of the nation or region and exchanged between them. The economic values of imports and exports (traded goods) may be adjusted upwards to take account of the value, especially scarcity value, of foreign exchange.

The other economic numéraire sometimes used is the *foreign exchange numéraire*, also called the *world price level*, which is based on a convertible foreign currency and on world or border prices (although the numerical values are usually expressed in local currency terms using an official exchange rate).[12] Imports and exports are valued at the border in foreign exchange terms, and goods not imported or exported (*non-traded goods*) may have their economic prices adjusted downwards to conform to this numéraire. Rather than distinguishing traded from non-traded goods, it is often better to distinguish *tradeable* from *non-tradeable* ones (see also Section 2.6).

In a country with significant trade or foreign exchange restrictions, the two numéraires cause costs and benefits to be given different numerical values, even if expressed in the same currency. The ratio between the two sets of prices is roughly the same for nearly all costs and benefits, so that prices expressed in one numéraire can be easily converted, approximately at least, to the other.

In most countries with convertible currencies the domestic pricing system is normally used, and it is usually the appropriate one for projects whose inputs and outputs are mostly produced and enjoyed within one country. Often the distinction is not mentioned at all in reporting an analysis. For the UK, *The Green Book* (H.M. Treasury, 2003) describes only the domestic pricing system. In countries where international trade is restricted and distorted, or the local currency is overvalued so that there are significant differences between the official exchange rate and a more realistic or a black market rate, the foreign exchange numéraire is sometimes preferred. It has some advantages when the analyst wishes to make specific adjustments for particular goods or classes of goods.

International funding agencies like the World Bank have in the past used both numéraires on different occasions. (Publications from or associated with the World Bank, up to the late 1980s, often recommended the foreign exchange numéraire, but the 1996 draft manual, and its successor (Belli *et al.*, 2001), unequivocally advocate the domestic pricing one. The 1997 guidelines of the Asian Development Bank do likewise, for most cases, although describing both.) Apart from some of the examples in Chapter 5, this book uses the domestic pricing numéraire, so as to minimise confusion and concentrate on shadow pricing for the correction of market distortions. The domestic pricing numéraire is the easier one for most people to identify with, and is recommended unless there is a strong reason (e.g. a client's or financier's insistence) to use the other one. It is important in conducting any CBA in an international setting to state clearly which numéraire is being used.

Many readers of this book will, in practice, never be concerned with the distinction, but those who need to know more, or who wish to be guided to the literature, will find more information in Appendix B.

Alongside the CBA methods described here, there is another system, called the effects method. It is a special kind of CBA, sometimes regarded as an alternative or even competitor to CBA, which implies a particular numéraire of its own. It was developed in France within a planning and national accounting system that provides coefficients describing interactions between parts of an economy, and it uses these to take account of indirect as well as direct project impacts on a national economy. In its usual form it uses financial prices without adjustment for market distortions, so it is not an economic analysis in the sense used in this book. The method is seldom used outside a few French-speaking countries. It is described in Appendix G.

2.4   Shadow pricing: practical application

The identity of the entity or group of people on whose behalf a project is analysed determines not only what should be valued but also how it should be valued. In a financial analysis this is simple enough: all ingoings and outgoings pertaining to the project are valued at what the financially relevant individual, firm or institution pays or is paid.

In an economic (or social) CBA the way in which costs and benefits should be valued is not so straightforward, as some of them may have no market and thus no financial prices, while others may have prices that, for instance because of market distortions, do not reflect their true value to the defined group of people. Special valuation techniques must be used, which go by the general name of *shadow pricing*. There are some general principles, but the precise methods often vary according to the type of cost or benefit involved. In this section some of the more common methods are described, starting with the simplest. The more specialised techniques are explained in more detail in Chapter 4 and the appendices. This section talks mainly about economic analysis, but most of the techniques apply also to the rare case of social CBA, some of the differences being explained in Appendix D.

For practical purposes, the usual method of applying economic prices is to start with the financial prices and then separate the cost estimate into categories and apply to each category a factor that converts a value in financial prices into a value in economic prices, the latter being some-times called *shadow prices* or *accounting prices*. The factor is called a *shadow price factor* or *SPF*.[13] How the SPFs are arrived at is the subject of Section 2.6 and Appendix C; their practical application is described here, and the underlying rationale is discussed in the next section, along with the concept of opportunity cost.

One special category is, of course, the transfer payments, such as taxes, which represent no significant economic cost despite their very real financial cost, and are to be omitted from an economic analysis. The simple computational way of handling this is to assign them an SPF of zero. Table 2.2 shows a simple example of shadow pricing, using some typical categories and SPFs for a civil engineering project with a significant unskilled labour component in a country with severe unemployment among unskilled people, and with a shortage of foreign exchange causing a scarcity of imported goods and services.

It can be seen that the shadow pricing in Table 2.2 gives to some cost categories a higher numerical value in economic terms than in financial terms, and to others a lower one. In this example the total economic cost is a smaller number than the total financial cost, because the low SPF for unskilled labour and the omission of transfer payments has outweighed the high SPFs for imported materials, skilled labour and machine hours; if the proportions of unskilled labour and taxes had

Table 2.2 *Example of shadow pricing of construction costs*

Cost category	Financial cost estimate[*]	Shadow price factor (SPF)	Economic cost estimate[*]
Imported materials	20	1.3	26
Other materials	10	1.0	10
Skilled labour	10	1.1	11
Unskilled labour	30	0.6	18
Machine usage	15	1.2	18
Taxes and duties	15	0	0
Other costs	20	1.0	20
Total	120		103

[*] Thousand pesos at 2016 financial prices.

been considerably smaller the economic cost might have turned out to be the higher number. It is important to note that the economic cost is of a different nature from the financial cost, and is measured in different units. For convenience, we use a monetary currency so that some of the SPFs will be unity, but this must not obscure the fact that the numbers in the right-hand column represent values to society in general (or whatever group has been defined), and not money.

2.5   Rationale of shadow pricing

As mentioned in Chapter 1, the prices obtaining in a free market are, in theory, indicators of people's *willingness-to-pay* for different goods, and are thus the correct measure of value to society. (The theory of the free market defines what classical economics calls the 'efficient' or 'optimal' allocation of resources. Free market prices are signals between producers and consumers, which, in theory at least, mediate the optimisation.) Real markets are, of course, different: perfect competition and perfect information are not achieved, and governments (acting more or less democratically on behalf of whole societies) distort prices for all sorts of good and bad reasons. When goods or services, appearing in a CBA as costs or as benefits, are traded in a relatively free market with few distortions, the market prices are a relatively good guide to willingness-to-pay, and are usually accepted and used in an economic CBA as they stand. When they are not so traded, the economic price can either be estimated directly, by finding out what people would be willing to pay if there were an effective market, or by taking the distorted real market price and adjusting it by shadow pricing.[14]

Although the rationale of economic or shadow pricing is discussed more fully in Appendix C, it can often be explained in terms of scarcity. In the example given in Table 2.2, the defined group of people, i.e. the nation, suffers a shortage of foreign exchange, and therefore of imported materials and machines. There is also a slight shortage of skilled labour, and so these three categories have SPFs greater than one to reflect their scarcity value from the nation's point of view. If these SPFs are consistently used for comparing projects, or for comparing different construction methods for the same project, the high SPFs will bias the choices in favour of projects and methods that use little of these scarce resources, which is the desired policy objective. Similarly, the low SPF for unskilled labour will make labour-intensive projects and methods less costly in economic terms, and so decisions based on such analysis will tend to choose labour-intensive methods and thus alleviate unemployment.

Thus, in effect, whatever their rationale, SPFs act like instruments of policy. Some economists might object that pure efficiency pricing is merely the application of willingness-to-pay, which is inherently correct, but this position is itself as laden with value judgements as any conscious policy.

A more general application of the scarcity idea is the concept of *opportunity cost*, the principles of which are explained in Box 2.4. This concept can be used in estimating shadow prices; in fact it is often regarded as the fundamental definition of economic pricing. Economic

---

Box 2.4 The general principle of opportunity cost

Opportunity cost is a powerful concept with many applications. We all use it in daily decisions, but usually unconsciously.

The opportunity cost of something is what we have to forgo or give up in order to obtain it.

If I go to a distant friend's wedding on a working day, it costs me not only the travel costs but also the loss of a day's wages. If, on the other hand, he gets married at a weekend, the cost is only the travel costs plus the loss of some leisure time (although I may value my leisure time very highly). If a design engineer asks to attend a one-day training course, her boss will probably ask what she would otherwise have been doing on the day in question: if she is working on a rush job that has to be out by Friday, the opportunity cost of her day's absence may be unacceptably high, but if the office is a bit slack that week it will be low. The opportunity cost of the time I spend washing up tonight will depend partly on whether there is anything I enjoy on television at the time (and to minimise the opportunity cost I will look at the television listing before deciding when to wash up). The opportunity cost of the use of a telephone line at the weekend is lower than that of the same line in working hours, because at busy times its use by me means foregoing its use by someone else; the telephone company may recognise this in its tariffs, and offer a low rate at weekends.

Strictly speaking, the opportunity cost of using some resources in a project is the benefit forgone in the best alternative use, not just any alternative use.

The principle of opportunity cost can be applied in financial, economic and social analyses, and to both costs and benefits.

---

or shadow prices are sometimes called 'social opportunity costs', where *social* means collective as opposed to private.

In a rural area, for instance, the use of unskilled labour on road-building is related to its use in agriculture, and the latter is, of course, seasonal. At harvest time or at other peak times for agricultural work, the opportunity cost to the nation of using an unskilled labourer for a day on the road is the value of the forgone work he might have been doing in the fields, which is quite a high value. At some other season his next most productive activity might be of little or no value to the nation, so the opportunity cost of using his work on the road would be low. A thorough economic analysis of road-building projects should take this seasonal variation of the shadow price of labour into account, thus biasing decisions in favour of projects and methods that use off-peak unskilled rural labour. This will produce roads at the least cost in terms of other productive activities foregone, and will alleviate rural unemployment. The analysis can either be done by estimating the economic value of the labourer's agricultural work directly, such as from crop budgets, or somewhat more crudely by estimating the seasonally varying SPF: multiplying the standard wage (the financial price) by, say, 1.0 in the harvest season, 0.3 in the slack season, and perhaps 0.6 at intermediate times.

Although taxes, duties and subsidies are, by definition, transfer payments, and thus omitted from an economic CBA, their existence can sometimes give a hint or guide to the presence and nature of factors requiring shadow pricing. Part (but only part) of a government's reason for putting a heavy tax on tobacco is that its use causes costs to society, such as hospital beds, which are not entirely borne financially by the smokers. A similar argument can apply to the taxing of fuels such as petrol, which cause pollution and other costs beyond the cost of making and transporting the fuel. An analyst may first strip out the large tax element of the petrol price (as a transfer payment), and then increase the resulting low price (the production and distribution cost) by a large factor or addition to represent society's non-marketed costs such as road-building, emergency services or health impacts. If the resulting economic price happens to end up numerically close to the financial price that the analyst started with, this is not because she has included taxes in an economic price, but because she has concluded that the tax level chosen by the government closely matches the economic and social costs that are beyond the production and distribution cost. Customs and excise duties are sometimes imposed to reflect scarcities, and again it may happen that the analyst may increase the price to reflect the scarcity by about as much as he decreases it when removing the transfer payment. A tax is, however, not necessarily a good guide to shadow pricing.[15]

2.6   Valuation of particular kinds of costs and benefits

This section discusses a few kinds of costs and benefits and the ways in which they are valued, although the details of such valuations are treated more thoroughly in the examples given in Chapter 4 and the appendices.

2.6.1   Valuation of direct construction costs

The main cost of most engineering projects, especially civil engineering ones, is the once-off construction cost at the start of the project's life. The starting point for valuing this sort of cost is a cost-estimate of the ordinary financial sort, which is usually developed by breaking the construction down into items or tasks and estimating for each item a unit cost (or rate) and a quantity. The unit cost is what the project or its promoter is expected to pay someone, a supplier or contractor or employee, and the quantity is the analyst's best-estimate of how many units will be needed. (As stated above, contingency allowances should not be included.)

For a financial analysis, this is all that is needed. For an economic analysis, the next step is the shadow pricing, or conversion into appropriate prices for whichever sort of analysis is being conducted. The usual process is to apply SPFs as in Table 2.2, remembering, of course, that transfer payments are omitted (or given an SPF of zero, which has the same effect). Engineers and planners will usually be able to obtain SPF values from someone else, such as an economist, who analyses the particular costs and benefits, or from a policy-setting body such as the nation's treasury or ministry of planning. In analyses whose principal users will include a development bank, such as, for example, the World Bank, one can often obtain guidance from the bank as to what SPFs should be used. As such official bodies and banks are likely to compare CBAs done by different analysts, they will want them to be mutually compatible and comparable, and so this use of advice from others is not an evasion of duty on the part of the analyst but a sensible means of ensuring compatibility.

The practical consequence, for engineers in particular, is that the financial cost estimate needs to be broken down, preferably from the

31

start of the estimating process, into the categories that may later need to have different SPFs. It is much easier for an estimator, at the end of his work, to lump together two categories that turn out not to need separating, than to split a category after the totals for all parts of the project have been aggregated. So it is wise to consider, before estimating begins, what categories may have different SPFs, even if those factors are not yet fixed. At the very least, the estimator will need to separate the tax element. Sometimes the estimate is made without categories, and then a guess is made (or deduced from other projects) as to the percentage of the total that represents all taxes and duties together. This crude approach may be acceptable at very coarse levels of planning, but is not generally advisable.

It is also sometimes necessary to consider whether the average or the marginal price or cost of an input is relevant. These terms are described in Box 2.5. If the quantity of cement to be used in a project is small relative to the regional or national consumption of cement, and if the national cement production capacity is underutilised, it could be argued that the economic cost of the cement used in the project is only the marginal production cost, because no new cement production capacity will be needed. The marginal cost may be significantly lower than the average cost, for a commodity whose production involves relatively high fixed costs and low variable costs; in such a case the choice of which unit cost to use must be guided by a realistic assessment of what would happen if and when the project was implemented.

2.6.2 Valuation of marketable benefits
When a project's main outputs are sold, as with factories or toll bridges, the obvious starting point is the market price. For an economic analysis, the analyst then has to ask whether this price is a realistic estimate of the commodity's true value to the nation as a whole. If the output is sold and bought in an approximately free market, it may well be so, and the economic price can then be numerically equal to the financial one (i.e. the SPF can be 1.0). If, however, the market is significantly affected by subsidies or taxes, or by firms or agencies taking advantage of a monopoly position, the financial price may evidently be quite a bad guide to economic value. An economist can sometimes adjust the financial price and arrive at an economic price by identifying the distortions (especially the more obvious ones such as taxes or subsidies on the goods in question, or on related or competing goods) and correcting for them directly; in

Box 2.5 Marginal and average unit costs

An important concept in economics is that of marginal price, marginal unit cost or marginal revenue. The marginal unit cost of producing or using something is the increase in production costs when one more unit is produced or used. Suppose I am spending £4000 a year to own and run a car that does 20 000 km a year, this being made up of:

Time-related costs	Insurance and licence fee	£500/year
	Time-related depreciation	£1500/year
Distance-related costs	Distance-related depreciation, 20 000 km at £0.05/km	£1000/year
	Servicing and consumables	£400/year
	Fuel at 6 l/100 km and £0.50/l	£600/year

If I now change my habits and drive 25 000 km/year, the time-related costs will remain unchanged at £2000, and the distance-related costs will increase in proportion to distance, from £2000 to £2500. The average cost has been reduced slightly, from 20 to 18 p/km (4000/20 000 and 4500/25 000), while the marginal unit cost is evidently 10 p/km (an extra £500 for an extra 5000 km). In more general terminology, the time-related costs are called *fixed costs* and the distance-related ones *variable costs*. Even when total costs can only be approximately divided between these categories, the concept is a very useful one.

Graph (a) in Figure 2.2 uses this simple example to show how the marginal unit cost is the slope of the total cost curve, while the average cost is the slope of a line from that curve to the origin. Most situations are not as simple as this, and the marginal unit cost changes with quantity. Graph (b) shows the case (much beloved of economists) where economies of scale give way to diseconomies of scale, and so the average cost has a minimum at the point where it equals the marginal unit cost.

effect, he describes a hypothetical free market in the affected commodity and estimates what prices would prevail in such a market. In other cases it is easier to ignore the distorted financial market and estimate the economic price directly, as described in the next section for those commodities for which there is no market.

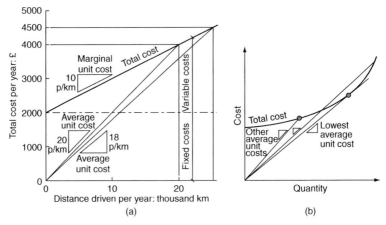

Figure 2.2 Marginal and average unit costs

Many projects have outputs that require special valuation procedures, e.g. the energy produced by a power station, the agricultural output from an irrigation scheme, the benefits of water supply or wastewater disposal, or those of road improvements. Some of these are described, with examples, in Chapter 4.

2.6.3   Valuation of non-marketed costs and benefits

Some project inputs and outputs have no financial price because they are not marketed and no one pays for them; examples are clean air, publicly accessible recreation areas or facilities, beautiful landscapes, and environmental impacts, both desirable and undesirable. In these cases one has to estimate the economic price directly, for instance by asking people what they would be willing to pay for a good if there were a market for it, or what compensation would just persuade them to do without another good. Alternatively, these willingnesses can be deduced indirectly by observing people's actions. In all these cases this is a job for a professional economist, perhaps a specialist in welfare economics, and is outside the scope of this book, although interested readers will find a discussion of the principles and some methods in Appendix C. However, all readers should be aware that the methods, which have names such as *contingent valuation* and *hedonic pricing*, are complicated and, even when well done, very approximate and liable to be controversial. Just going out into the street with a clipboard and asking people what they would be willing to pay for a bypass will not do: many people will have no idea, some will answer tactically, and all will be heavily influenced by the precise way in which the question is asked.

Some project inputs, notably voluntary work done for altruistic motives, and family labour in small farms or enterprises, are not paid for in the real market but could, in principle, have a market price. These must also be included in an economic analysis, being valued by reference to similar inputs that are paid for.

2.6.4 Depletion of finite resources

If a project uses, as an input, something that is taken from a finite stock, the economic cost is more than just the cost of extraction at economic prices. The extra amount, sometimes called a *depletion premium*, represents the extra cost that would be incurred in the future, after the present stock is exhausted, in using a more expensive source or a substitute because of the depletion caused by the project. Whether the use of the alternative source or substitute occurs in the project's lifetime or later, it still represents a cost to the nation caused by the project, and should be counted. The further in the future it lies, however, the less impact it has during any one year of the project's analysis period, and the future extra extraction costs are discounted back to each project year before being labelled depletion premium and added to the actual extraction costs (discounting is explained in Chapter 3). Thus the depletion premium, as a figure applied in a particular year, increases as the date of depletion approaches. Finite resources that a project may help to deplete include not only fossil fuels but also easily accessible clean water, topsoil, clean air, the ozone layer and endangered species.

In practice, the computation of a depletion premium usually requires information that can only be guessed at, such as the size of the stock of unexploited resource (the world's reserves of fossil fuels are constantly being re-estimated), and the cost of the likely substitute (which might involve technology not yet invented). The premium is often ignored in CBA. The recommended procedure for most cases is to mention the depletion of probably finite resources alongside a CBA, and perhaps to include it crudely in sensitivity tests, but not to attempt to value it in the base case. In a case where sensitivity tests or common sense indicate that the depletion may be crucial to the analysis, a valuation may then be required, although the details are beyond the scope of this book.[16]

2.6.5   Distribution between groups of people

In a social CBA it will sometimes be desired to adjust the valuation of costs or benefits according to which people they affect. Quite apart from ethical considerations, it is generally agreed that the marginal unit of consumption is more valuable to a poor person than to a rich person, and even for one person the seventeenth chocolate biscuit is usually not as highly appreciated as the first. (This is called the *diminishing marginal utility of consumption*.) For some analyses the groups of beneficiaries are identified and the benefits pertaining to each group are separately valued to take account of this. The details are complex and are considered further in Appendix D. Although advocated by Squire and van der Tak of the World Bank in 1975, the method that uses consumption levels is in fact very seldom applied, by the Bank or anyone else.[17] Explicit analysis of gainers and losers is more informative.

2.7   Double-counting

It is obvious, in principle, that each relevant kind of cost or benefit should be counted once and only once, but sometimes it is less easy to ensure this in practice. For example, an economic or social analysis may use shadow pricing of costs to take account of a project's effect on employment or rural incomes: to add amounts to the benefits because of these indirect effects would constitute double-counting. Special calculations of foreign exchange earnings or multiplier effects also run the danger of double-counting.[18] The antidote is common sense, and clear definitions of the project, the beneficiaries and the type of analysis.

A similar need for caution arises when CBA is used alongside other analyses and criteria to guide a decision, whether in a formal multicriterion analysis (see Appendix F) or not. If shadow pricing has taken into account some project effect such as employment generation, it may be wrong to quote that effect again alongside the CBA as if it were an additional benefit. However, it is acceptable to quote overlapping criteria (ones that are not independent of each other or have some common elements) in a multicriterion analysis that explicitly acknowledges the overlaps.

Example (for readers to work through)

Readers may wish to work through the example below as a way of reviewing the content of this chapter. Some questions will be easier to answer after reading Chapter 4, but all can be attempted at this stage. There are no unique correct answers to the questions, but a set of tentative model answers is given in Appendix H.

The analysis

The feasibility study of a water resources development project in a Latin American country requires an economic analysis. The project comprises a dam, a hydropower station and an irrigation project, all located in a previously almost uninhabited area. The reservoir formed by the dam would inundate a national highway, a short stretch of which would have to be rebuilt at a higher level. The reservoir would flood an area that has some visual merit as a landscape, but would itself be visually attractive and would provide water-sport facilities and some new wildlife habitats. The power station would meet the region's electricity demand for about 15 years, and would later provide convenient peak power in a grid with other power stations. Some of the construction could be carried out either by labour-intensive methods or by imported machinery. The country is a net importer of oil and has high unemployment among unskilled people. The irrigation project would produce wheat for local consumption, and tobacco which would replace imports: the market price of tobacco would fall and the citizens would smoke more.

The project is to be analysed on behalf of the government repre- senting the whole country, but is located in a remote region whose development the government wishes to promote by conscious policy measures. The government also wishes to favour choices that promote employment of unskilled people.

Questions

(a) What categories should the financial construction costs of the engineering works be broken down into, and roughly what levels of shadow price factor (SPF) would you expect for each category?
(b) Who would you consult in order to arrive at the final SPFs?
(c) What other costs would need to be quantified?
(d) What benefits would need to be quantified and valued?
(e) What significant project effects would you exclude from the CBA and mention to the decision-makers alongside it?
(f) Which parts of the analysis might be controversial, and thus need particular care in presentation when the feasibility report is written?

Notes

1 Rehabilitation initiatives are discussed in detail in Yaffey and Tribe (1992).

2 See also Belli *et al.* (2001, Ch. 3 and the example in Ch. 12). The Sila example is used again in Chapter 5 of this book.

3 This is discussed also in Belli *et al.* (2001).

4 A more extensive explanation is given in the World Bank handbook (Belli *et al.*, 2001, Ch. 4).

5 An example is given in Belli *et al.* (2001, p. 32).

6 Financing bodies such as the World Bank often receive requests for finance to finish projects that should perhaps never have been started, as in Box 2.2. The rule about excluding sunk costs still applies, and it may indeed be worthwhile to finish such a project, but the analyst has to be very careful about the definition of the without-project situation. Cancellation or truncation may be real options, and it may be necessary to analyse more than one of these alongside the option to complete the project; this is discussed in Yaffey and Tribe (1992, Ch. 5).

7 The World Bank handbook (Belli *et al.*, 2001) does not make the distinction, lumping categories (a) and (b) together under the name *physical contingencies*. Yaffey (1992) implicitly does the same, and even advocates inclusion of the allowance in CBA unless there is a risk analysis. The Overseas Development Administration (ODA, 1988, Ch. 6) makes the distinction as here, and says the true contingency should be excluded unless the analyst thinks it ought to be part of the best-estimate, i.e. reclassifies it. The British public sector appraisal guide argues, as here, that only the best-estimate should be included in the analysis (H.M. Treasury 2003, Sections 5.57 to 5.66).

8 A CBA that includes depreciation is presented in Zerbe and Dively (1994, but such a procedure is not recommended): The World Bank handbook is explicit on this (Belli *et al.*, 2001, p. 27).

9 Unless the term *contingencies* is being loosely used and really means miscellaneous items within the best-estimate, as discussed in Section 2.2.5.

10 Unless a before-financing analysis is wanted.

11 Other names often used are the willingness-to-pay numéraire and the United Nations Industrial Development Organisation (UNIDO) system, the latter referring to a UNIDO publication of 1972. The precise definition is that all costs and benefits are valued in terms of their impacts on the society's consumption, measured as what people are willing to pay for each good or service (at the margin, i.e. the willingness-to-pay for one more unit; see Box 2.5).

12 Strictly defined as free foreign exchange in the hands of the government of the nation for whom the analysis is being done; a sort of national opportunity cost. This numéraire is sometimes called the Organisation for Economic Co-operation and Development (OECD) one, from the

organisation which produced a manual in 1968; it is also associated with the names of that manual's authors, Little and Mirrlees, and sometimes also with Squire and van der Tak; hence it is often referred to as the LMST numéraire. The World Bank handbook calls it the 'border price level', whether expressed in local or foreign currency.

13 This factor is sometimes called a 'conversion factor' (e.g. in the World Bank handbook (Belli *et al.*, 2001)), but that term is more usually reserved for a factor related to the different numéraires described in Appendix B. Brent (2006) calls it the 'accounting ratio'.

14 The World Bank handbook emphasises that the market distortions, once they have been analysed by an economist for the purpose of adjusting financial prices to economic ones, are also very revealing about the distribution of costs and benefits between parties, and about a government's policies (Belli *et al.*, 2001, p. 15).

15 See Hanley and Spash (1993, Section 1.3, pp. 10–11).

16 When there is significant exploitation of depletable resources, it may be necessary to compute a 'depletion premium', as set out in Appendix 6 of Asian Development Bank (1997).

17 See Squire and van der Tak (1975, Ch. 7), Brent (1990, 1998) and the pair of articles about it, by Brent and by Weiss, in *Project Appraisal* (1994). The World Bank handbook gives very little emphasis to distributional weights, preferring to record the distribution of costs and benefits between affected parties (gainers and losers) as a descriptive exercise outside the calculation of the CBA indicator (Belli *et al.*, 2001).

18 Squire and van der Tak (1975, p. 24).

3

Resources and time: discounting and indicators

But wherefore do not you a mightier way
Make war upon this bloody tyrant, Time?[1]

3.1   Time preference

Most of us prefer jam today to jam tomorrow. When considering using resources or enjoying benefits, we are not indifferent to timing. Reasons and justifications for this are many and disputed, and some are discussed below, but for the moment let us accept the fact and examine its implications for CBA.

The name of this fact is *time preference*, and it begins with individual time preference. To avoid confusion with monetary ideas such as inflation and interest, and to concentrate on resources in a general sense, I will begin by saying that I personally would just as rather receive 100 jars of jam today as 110 jars of jam one year hence: jars of jam are a proxy for resources in general, and those numbers happen to describe my personal time preference rate. If you make me a credible offer of 111 jars of jam next year in exchange for 100 jars that I have right now, I will accept because that is one jar more than what I would regard as equivalent, but if you offer me 109 jars a year hence I will prefer to keep the 100 jars that I have. In the jargon of economists, I am *indifferent* between 100 jars now and 110 next year.

My reasons might be mixed. To give advance notice of some that will be discussed in Section 3.9, I might just be irrationally or irresponsibly impatient; I might fear not to be alive next year to eat jam; I might expect to be richer next year and therefore less urgently interested in jam; or I might want to sell some of my 100 jars, put the money in a good investment account, and with the interest buy considerably more jam in the future.

Assuming I am consistent and do not change my attitude to time and jam, I will have the same ratio next year, so that 100 jars next year will have the same value to me as 110 jars the year after. This means that

Cost–benefit analysis – A practical guide
ISBN: 978-0-7277-4134-9

121 jars in 2019 has the same value to me as 110 jars in 2018, or 100 jars in 2017. All three packages have the same value to me, where a package is defined by three bits of information: a number (110), a unit (jars of jam) and a point in time (the year 2018). In CBA it is conventional to designate one point in time as year 0, so that in this example 2017 is year 0, 2018 is year 1 and 2019 is year 2. Among statements about resource packages, any referring to year 0 is called a *present value* (PV), and any referring to a later year is called a *future value* (FV). This defines an arbitrary 'relative calendar' for calculations of present value. How it is related to a project's timing is a separate matter: year 0 may be the year of decision, the year before the first costs or benefits fall, or some other year. Negative year numbers are possible, but tend to cause confusion.

Given a statement about the value of a quantified package of resources, or costs or benefits, seen from one moment in time, it is useful to be able to make a statement about its value as seen from some other moment. I can say that the package '121 jars of jam in year 2' has (at my time preference rate) the same value as 100 jars in year 0. Or that a jar of jam has less value when seen from a moment two years earlier, by the ratio 100/121. Or that an FV in year 2 of 121 has an equivalent present value of 100 (present value implies year 0).

The numbers of units (jars) are referred to an FV and PV, and FV is the larger number. (When we are comparing equivalent packages, the numbers get bigger as the unit of value gets smaller; I value 2019 jam less, so I need more jars of it to be equivalent to a certain amount of 2017 jam.) The ratio of PV to FV in this example is:

$$PV/FV_{(year\ 1)} = 1/1.10$$

In the case of the package two years ahead:

$$PV/FV_{(year\ 2)} = 1/1.21$$

In general, for any number of years:

$$PV/FV_{(year\ n)} = 1/(1 + r)^n$$

where r is my annual time preference rate expressed as a fraction, in this case 0.10; it is often expressed as a percentage.

This is, of course, the same arithmetic as compound interest computations, but applied to a different purpose. People are often surprised at the large ratios implied by quite moderate time preference rates. For instance, with this rate of 0.10, or 10%, over a period of 20 years, the ratio PV/FV is only 0.149, or about a seventh. This means that it would take the prospect of seven jars of jam 20 years hence to weigh as heavily on my judgement as a single jar today. Yet, for better or worse, 10% is a rate commonly used in guiding decisions.

Before this concept can be used in CBA, a few more assumptions have to be made, although they are seldom mentioned or discussed. The first is that the same time preference rate applies to different sorts of things, so the same rate can be applied to jam and to concrete. The second is that there is such a thing as a collective time preference rate, often called the *social time preference rate*, similar to my personal one. The third assumption is that it is possible to estimate such a rate and sensible to use it in guiding decisions. For the moment we will blithely make all these assumptions, although they will be examined again later (in Section 3.9 and Appendix E). The rate used in a CBA, whatever its justification, is called the *discount rate*. In this book the time unit will always be taken to be a year. Conventionally, the moment when a certain amount of money or resources is used or produced is merely referred to as a particular year. This implies that, as a reasonable approximation, all the year's activity is lumped in one moment, such as its midpoint. One can distinguish between mid-year and end-year moments (the discounting ratio for half a year is, of course, the square root of $1/(1 + r)$), or even between months, but the rest of the data are seldom accurate enough to make this refinement worthwhile.

In the next section the arithmetical procedures of discounting will be described. They are confusingly similar to financial calculations such as compound interest and annuities, and many books and manuals increase the confusion by treating them as identical. This book deliberately introduces discounting as a matter of time and resources, not a matter primarily about money.

3.2   Discounting

Discounting is the arithmetical process of converting value statements referring to one moment in time to their equivalent value statements referring to another moment in time. Normally it refers to finding the PV of an FV or a set of several FVs. It uses the above relationship expressing the ratio of PV to FV in terms of the discount rate r and the time difference n. One can calculate PVs directly using the formula, but there are three easier ways:

- tables of factors for particular values of n and r
- functions in computer software, especially spreadsheets
- special function keys on some calculators (increasingly rare as the computer method becomes more widely used).

Table 3.1 Part of a discounting table: 10%/year discount rate

No. of years (n)	PV/FV	PV/RV
1	0.9091	0.9091
2	0.8264	1.7355
3	0.7513	2.4869
4	0.6830	3.1699
⋮	⋮	⋮
10	0.3855	6.1446
⋮	⋮	⋮
20	0.1486	8.5136
⋮	⋮	⋮
50	0.00852	9.9148
⋮	⋮	⋮
100	0.00007	9.9993

Discounting tables are given in Appendix J and this section shows how to use them, while computer functions are explained in Box 3.3. An excerpt from the appendix, illustrating part of the table for $r = 10\%$, is reproduced in Table 3.1. This table, one of several for various discount rates, has three columns, the first for the time difference in years (n), the second for the ratio PV/FV defined by the above formula, and the third for another ratio called PV/RV, which will be explained shortly. The simplest use of the table is to convert a single FV (a value statement referring to a future moment, namely year n) into the equivalent PV (a value statement referring to year 0). If, for instance, the FV is 121 units (jars, dollars, litres or whatever) in year 2, and the discount rate is 10%, we find the relevant PV/FV ratio, from Table 3.1 or Appendix J, to be 0.8264. We multiply the numerical FV by that factor to get the PV: $121 \times 0.8264 = 100$. So the PV is 100. (This is, of course, the example used in the previous section and the PV/FV factor 0.8264 is $1/1.121$ or $1/1.10^2$ or $1/(1 + r)^n$.) If the FV were 300 units in year 4, the PV would be $300 \times 0.6830 = 204.9$ units. If the FV were 573 units in year 20 the PV would be $573 \times 0.1486 = 85.1$ units.

The significance of PVs is that they are all expressed as values seen from the same moment, so they are in a common unit and can be added together. If a project produced benefits quantified as 121 units in year 2, 300 units in year 4, and 573 units in year 20, the PV of all benefits would be $100 + 204.9 + 85.1 = 390$ units. The reader who wishes to practise this simple discounting operation, using the tables in Appendix J, will find an example in Box 3.1.

> ### Box 3.1 Reader's worked example on simple discounting
>
> Find the total present value of the following series of *future* values, if the discount rate is 7%.
>
> | Year | FV |
> |------|-----|
> | 1 | 314 |
> | 2 | 291 |
> | 3 | 109 |
> | 4 | 86 |

It often happens that we need to discount a series of numerically equal FVs in successive future years, for instance the steady benefits of a project after construction is finished, or its steady operating costs. We could, of course, write down the series of FVs and discount each one separately. For instance, to discount (at a rate of 10%) a series of FVs each equal to 56 and running from year 1 to year 4, we could multiply the year 1 value of 56 by 0.9091 and write down 50.91, then multiply the year 2 value of 56 by 0.8264 and write 46.28, and so on. Having done this four times we could add up the four PVs (50.91, 46.28, 42.07 and 38.25) and get the total PV of the series, namely 177.51 units. This would be correct, but tedious, and most of us would soon choose the short cut of adding the PV/FV factors first and then multiplying their sum, 3.1698, by the common FV number 56, arriving directly at the total PV of 177.51. The tables in Appendix J, like the excerpt in Table 3.1, make this easier by tabulating the sums of PV/FV factors in the column headed PV/RV, where RV stands for *repeated value*. In this case the number of years is 4 and the RV is 56, so from the table we can obtain the factor PV/RV = 3.1699 directly, and multiply it by RV (56) to get the total PV (177.51) in a single step. (The tables are correct to four places of decimals, and the value 3.1699 is more accurate than the apparent sum of the four PV/FV factors, 3.1698, although the difference has, of course, no importance in practice. The factors PV/RV are merely the cumulative sums of the PV/FV numbers beside and above them in each table, as can be seen in any of the tables in Appendix J. The formula for PV/RV is:

$$\frac{1}{r}\left\{1 - \frac{1}{(1+r)^n}\right\}$$

which for long discounting periods tends to $1/r$, and thus the ratio for a series to perpetuity is $1/r$.)[2]

The PV/RV factors given in tables always apply to a series of equal FVs starting in year 1. We often want to discount a series starting in some later year. The simple way to do this is to take the factor starting in year 1 and ending when the desired series ends, and then subtract the factor starting in year 1 and ending the year *before* the desired series starts. For instance, suppose we want the PV of a series of FVs each equal to 73 units, running from year 5 to year 12, at a discount rate of 8%/year. The PV of a series of 73s from year 1 to year 12 would be 73×7.5361 (using the tables in Appendix J). The PV of a similar series from year 1 to year 4 would be 73×3.3121. The PV of the desired series (73s from year 5 to year 12) is clearly the difference between these two, or

$$73 \times (7.5361 - 3.3121) = 73 \times 4.2240 = 308 \text{ units}$$

Discounting each year's value individually using PV/FV factors, and then adding the eight PVs, would give the same result but would take longer. It is easy, when in a hurry, to make the mistake of subtracting the factor for year 5, because the desired series begins in year 5, but in fact the subtracted series must, of course, end the year *before* the desired one begins. Readers who wish to practise this procedure can do so by completing the worked example in Box 3.2, which builds on the one given in Box 3.1.

Discounting deals with the time value of resources or effects of any kind. Usually it is applied to money or to an economic or social value that is expressed in monetary units by means of some valuation technique. It can, however, be applied to anything whose value needs to

---

Box 3.2 Reader's worked example with repeating values

Find the total present value of the following series of future values, if the discount rate is 7%.

Year	FV
1	314
2	291
3	109
4	86
5–20	7 each year

be assessed at different times. For instance, it is common to discount the annual volumes of water delivered by a water supply scheme, so as to produce a *discounted total volume* that can be compared with a discounted total cost, in a cost-effectiveness analysis. Similarly, one can discount numbers of patients treated at a hospital, or numbers of children taught in a school, or tonnes of coal carried by a railway.

3.3   Indicators and their calculation

Discounting can be used to calculate numbers called *indicators*, which describe the relative merit or desirability of projects or options being studied. The most basic are the *net present value* (NPV) and the *benefit/cost ratio* (B/C). The starting-point for both is to list the costs and the benefits by year, and to discount both series. This gives two numbers: the total discounted costs C, and the total discounted benefits B. Then NPV is the difference between them, B – C, and the benefit/cost ratio is the ratio between them, B/C. The word *net* in this context means the difference between benefits and costs, and it can be applied either to a particular year or to a pair of PVs derived by discounting. Table 3.2 gives an example of the costs and benefits, listed by year, for a hypothetical project. Its costs consist of heavy initial capital expenditure on construction of the project works in years 1–4, and then a steady running cost of 7 units/year from year 5 to the end of the *analysis period*, assumed to be the end of the project's life, in year 20. Its benefits when it is running at full capacity are 200 units/year, but they do not start instantly at that level; instead, the project begins to produce benefits, from the part of its infrastructure that has been completed first, in year 4, and builds up over five years as the whole of the project comes into operation and the participants master the new technology. (Readers who have worked through the example in Boxes 3.1 and 3.2 will find that they have already discounted the cost stream of Table 3.2; if they require further practice they can discount the benefits and/or the net benefit stream.)

Table 3.2 shows the net benefit for each year, and the discounted value of each of the three streams, the benefits, the costs and the net benefits. The discount rate in this example is 7%/year: at this rate the total discounted benefit B amounts to 1230 units, and the total discounted cost C to 753 units. The NPV at a 7% discount rate is therefore +477 units (= 1230 – 753), and the B/C ratio at a 7% discount rate is 1.63 (1230/753). If we discount the net benefit stream in the fourth column in Table 3.2 the result is 477 units: the

Table 3.2 Discounting and indicators for a hypothetical project: discount rate 7%/year

Year	Benefits	Costs	Net benefits
1		314	−314
2		291	−291
3		109	−109
4	10	86	−76
5	30	7	23
6	90	7	83
7	160	7	23
8	200	7	83
9	200	7	153
10	200	7	193
11	200	7	193
12	200	7	193
13	200	7	193
14	200	7	193
15	200	7	193
16	200	7	193
17	200	7	193
18	200	7	193
19	200	7	193
20	200	7	193
PVs	**B**	**C**	**NPV**
	1230	753	477

B/C = 1.63

NPV is the same whether we do the *net* part of the calculation (net benefit = benefit − cost) first or the *present* (discounting) part first.

It is interesting to observe the effect of different discount rates on this example. If the discounting, instead of being done at a rate of 7%, were done successively at 10%, 12%, 15% and 20%, the results, expressed in units of PV, would be as shown in Table 3.3. (All these discount rates are per year and one should strictly write 7%/year etc., but as it is very rare to use any other time period it is conventional to write merely 7%.)

Because this project has its costs concentrated in early years and its benefits spread mainly over later years, the increasing discount rate, while reducing the numerical value of both the total discounted benefit and the total discounted cost, always reduces the discounted benefit faster than it reduces the discounted cost. Thus an increasing discount rate leads to a decreasing NPV and a decreasing B/C ratio. There is evidently some particular discount rate that will make the total

Table 3.3 Discounting at 10%, 12%, 15% and 20% in the example given in Table 3.2

Discount rate	Total discounted benefit (B)	Total discounted cost (C)	NPV	B/C ratio
7	1230	753	+477	1.63
10	887	704	+83	1.26
12	722	676	+47	1.06
15	539	638	−98	0.84
20	345	584	−240	0.58

discounted benefit equal to the total discounted cost, so that for that discount rate the NPV will be zero and the B/C ratio will be one. This particular discount rate is called the *internal rate of return* (IRR). If the analysis is an economic one it is the *economic* IRR (EIRR or ERR), and if the analysis is a financial one it is the *financial* one (FIRR).

In the above example it is evident from the tabulated figures that the IRR lies between 12%, which gives a positive NPV of +47 units, and 15%, which gives a negative NPV of −98. It is also evident that it is nearer to 12% than to 15%. If one wished to find its value one could proceed by trial and error, beginning by calculating the NPV for 13%, which turns out to be small and negative (−8.3 units). This shows that the IRR must be between 12% and 13%, and nearer the latter. Although the function is not linear, it is a reasonably smooth one, and within a range of one or two percentage units a linear or graphical interpolation is a good approximation. In this case linear interpolation between 12% and 13% indicates an IRR of 12.85%, whereas the true value, as further trial and error would show, is 12.84%: the difference is, of course, trivial. It is possible, but very rare in practice, for a project to have a net benefit stream (cash flow) that starts negative, turns positive, then crosses the axis again: this can give rise to two or more discount rates with NPV = 0, and thus to multiple IRRs.[3] It is also not unknown for a project to have no IRR at all, because the benefits are high and the costs late. If the net benefit stream has either of these shapes, the analyst should plot it graphically and ensure that any and all IRRs are found; the result should be reported candidly and interpreted with caution, preferably relying mainly on other indicators. In the multiple IRR case, a computer spreadsheet will generally converge on only one value, which one being dependent on where the iteration started (Box 3.3).

Thus the three indicators NPV, B/C and IRR are arrived at by discounting. They apply equally to financial, economic and social

48

CBAs, according to the decision to be guided by each analysis. Which indicator is appropriate depends on the kind of decision being guided, as discussed in Section 3.6. Although it is useful to be able to calculate indicators using tables and a pocket calculator, as the reader who has worked through the above examples can do, most analysts are more likely to use a computer. Although the job can be done through a programming language such as FORTRAN, it is far more convenient to use a spreadsheet (a self-calculating table made with a commercially available software package). This is easy to do, and is explained in Box 3.3. Spreadsheets are especially convenient because, although it may take a while to set one up for a particular analysis, once that is done the inputs can be changed and the results recalculated and presented in a matter of seconds. This enables the analysis to be worked through early in a study with preliminary estimates of many data inputs, to give advance warning of likely outcomes. It makes sensitivity tests (described in Section 3.7) especially quick and easy, and thus encourages their use. If a spreadsheet is neatly laid out, which is a good discipline anyway to avoid confusion and errors, it can be printed and used directly in a report, so it both performs and reports the calculation in one step, saving time and avoiding the need for proofreading to guard against copying errors. An example of a spreadsheet used in this way is given in Appendix I.

The NPV and B/C values depend on the discount rate, and so the rate must be stated when these values are quoted; for example, 'The NPV is 10.7 million euros at a discount rate of 9%'. The EIRR (or any IRR) is, however, independent of any assumed discount rate, as it is itself a particular discount rate.

Alone among the three indicators discussed so far, the numerical value of the B/C ratio, and the consequences of its use for any purpose, depends on precisely how benefits and costs are defined. Some effects of a project or decision can reasonably be treated either as positive benefits or as negative costs. For instance, in a project to restore and stabilise an eroding beach, although the dominant benefit was the future enjoyment of the beach for bathing, there was also expected to be a dramatic reduction in the annual cost of repairing the coastal road, which tended to be damaged by waves.[4] The annual reduction in road maintenance cost, after appropriate shadow pricing, could either be added to the benefit stream or subtracted from the cost stream. The choice would make no difference to NPV, because NPV concerns net benefits (benefit − cost) anyway, and would not affect the IRR because that is defined by NPV = 0. The B/C ratio of any project or option

49

Box 3.3 *Discounting in computer spreadsheets*

One can insert the basic discount formula of Section 3.1 in a spreadsheet or workbook, but it is usually more convenient to let the spreadsheet software accomplish discounting automatically by specifying the appropriate function. A function in this context is a coded instruction that tells the computer to do a certain calculation using certain numbers as input. The user goes to the spreadsheet cell where he wants the result to appear and types in the name of the function followed by arguments enclosed by brackets and separated by commas; the arguments either contain the input numbers or tell the computer where to find them by means of cell references to other places in the spreadsheet where those numbers are already held.

The spreadsheet software Excel serves here as an example. It has two relevant functions, the names of which are NPV and IRR. NPV produces the PV of a list of values, whether equal or not. This function has the format 'NPV(D, range)', where the first argument D, is the discount rate as a ratio (0.10 for 10%), and the second is a range, such as 'B3:B22', within which the list of numbers to be discounted is located. The computer will assume that the first (leftmost or uppermost) number in the list belongs to year 1, and so if a stream of costs or benefits starts in a later year it is best to set up and refer to a range with zeros entered for the earlier years, from year 1. In Excel one must never leave a cell within the range blank, as this will lead to an incorrect result, even if the range was specified correctly.

The discount rate can be entered in the formula directly as a number, such as 0.1, but it is usually better to refer to a cell elsewhere, which contains the rate, so that the rate will be printed in the spreadsheet and can be changed at any time without having to edit the function. For instance, the benefit stream in the example in Table 3.2 is discounted by arranging the discount rates, as percentages, in cell C3, and the benefit stream in cells B8 (for year 1) to B27 (for year 20); then the formula in cell B30, to produce the discounted benefit, is 'NPV(C3/100, B8:B27)'.

The other function has the format 'IRR(G, range)', and produces the IRR of a series of values in the named range. Again the range must not contain any blank cells, so the cell for a year with no value must contain a zero. The computer finds the IRR by internal trial and error, and needs to start the process somewhere, so the user has to insert a guessed trial IRR value G to start the automatic iterative solution. It usually helps to insert a guess above the likely result rather than below it. If the iteration fails to converge, the computer will return an error message, in which case the user merely needs to try a different starter value G. The very unusual multiple IRR case requires special care, as different starter guesses may produce different IRR results.

There is a third function called PV to discount a stream of equal values, but this is usually better done with a column of numbers and 'NPV'. Most other spreadsheet packages have similar discounting functions, and sometimes others as well, often with confusing names. Some packages are more tolerant of blank cells within ranges than Excel is.

would, however, be different according to how this item was classified, except for options that happened to have a B/C ratio of unity: the further from unity the more marked the difference. Similarly, in analysing agricultural projects, the cost of fertilisers is often subtracted from gross crop value before project benefits are computed, so that it is treated as a negative benefit rather than as a cost. The fact that the value of a B/C ratio depends on how such elements are treated is a practical disadvantage of this indicator for any use involving comparisons between projects or between options: before making a comparison one needs to ensure that all the things being compared have their B/C ratios defined in identical ways, or at least in compatible ways. As, when it comes to detail, there are about as many B/C definitions as there are analysts, this can be difficult.

There is another class of indicators that are here regarded as a subset of B/C ratios, although some people regard them as a separate set. Occasionally called 'limited factor analysis', they are indicators that explicitly relate the project's desired effect to its use of the key limited resource, such as investment funds. To most decision-makers this seems intuitively a sensible approach. The main indicator in this class is the *net benefit/investment ratio*,[5] usually abbreviated as N/K, which in general means the PV of net benefits, this time meaning gross benefits minus all costs except investment costs, divided by investment costs. (In a financial analysis this is the monetary return per unit of investment, which is an indicator often used to guide business decisions.) Again, there is room for varying definitions and borderline categories, such as what is and is not an investment cost. Some analysts put the PV of the net benefit stream after that stream has turned positive on the top of the fraction, and the PV of the earlier years before it turns positive on the bottom. This is simple and unambiguous, provided the net benefit stream turns positive only once (like the well-behaved example in Table 3.2), and decisively, rather than wobbling about near zero. If there are occasional later years with negative net benefits, for instance when replacement costs fall, ambiguity can still be avoided by rewording the definition as 'after/before the stream *first* turns positive'. If, however, some or all of the projects being compared have significant late costs, such as the decommissioning costs of a nuclear power station, this definition is likely to be ambiguous and inappropriate. Other analysts define the bottom of the fraction more functionally, by considering the nature of the limiting resource, generally investment funds under the control of the decision-maker (within a defined period): this may be arbitrary but it is closer to the spirit of limited factor analysis.

This section has defined the three indicators NPV, B/C and IRR, together with some variants, and shown how to calculate them in practice. The next two sections deal with matters that tend to arise in the process of calculating indicators, and the question of which one is appropriate for a particular purpose is covered in Section 3.6.

3.4   Length of analysis period

For each analysis a choice has to be made of the period to be analysed in the discounting procedure. In principle this is related to the expected lifetime of the project being analysed, or the period over which the forthcoming decision will take effect (e.g. up to the moment after which the with-project and without-project situations will be, for practical purposes, the same). In practice this principle may not be much help, as many decisions and projects, especially civil engineering ones, have effects that last almost indefinitely. With effective maintenance, and perhaps occasional refurbishment (the cost of both of which can, and usually should, be counted in the CBA), many bridges, tunnels, roads, harbours, canals and such-like will serve for centuries, although their use might cease earlier for commercial reasons. A pragmatic way to settle this choice for most CBAs is to note that, unless the discount rate is very low, costs and benefits in the distant future will make little difference to the outcome of the analysis. At the common discount rate of 10%, a stream of net benefits valued at £1 million per year will give a PV of £9.43 million if discounted for the next 30 years, while the corresponding figure for discounting over 40 years is 9.78, that for 50 years 9.91, and that for 100 years 9.999, or £10 million if discounted indefinitely (see Appendix J, PV/RV factors for 10%).

As the estimates of annual net benefits usually contain uncertainties at least of the order of 5%, the accuracy of the analysis is barely affected if the discounting is terminated after 30 or 40 years, even if the benefits really continue for much longer. This argument can be used to justify a truncated analysis, provided that the nature of the decision or project is not such that drastically larger costs or benefits are likely to crop up in later decades. The classic example of the other case is, of course, a nuclear power station, where massive decommissioning and waste disposal costs are expected at, and long after, the end of the plant's productive life. In such cases, or when (for whatever reason) a low discount rate is used, it is better to use a long analysis period so as to avoid distortions and criticism. (This can be done without extending the discounting of *all* costs and benefits, by estimating a lumped cost in the last year of analysis that represents the environmental costs after that date, discounted to that date as if it were year 0. It would, of course, be inconsistent to compute such a lumped value using one discount rate and then carry out the main analysis with another, which the use of the IRR would usually do.)

Another argument for truncating an analysis period is that the forecasting of benefits into the distant future is often particularly uncertain: a navigation canal may well last for centuries but the benefits may almost cease after a few decades if technological advance renders water transport obsolete. The same could happen to many other sorts of technology, and it may be prudent not to claim benefits that are subject to such uncertainty, although this amounts to a lower-bound approach and may not be appropriate for comparing dissimilar projects, between which it would introduce a bias.

Analysts sometimes terminate the discounting after relatively few years and then allow a nominal lump-sum benefit in the last year to account for the *residual value* of project infrastructure at that moment. This is a messy way of dealing with the matter. The value at such a time depends on the net benefits that the residual infrastructure could produce in the further future, and those net benefits would need to be discounted to the date of the residual value figure: the figure would, of course, depend on the discount rate used, and so assuming any figure independently of the choice of discount rate for the main analysis would be inconsistent. Now that computers make the discounting of long analysis periods relatively painless, it is better and

53

easier to extend the period far enough to be able to convince everyone that any remaining residual value is arithmetically negligible.

Some kinds of project have predictable lumped future costs that are distinct from normal annual maintenance or other routine recurrent costs. A common example is the mechanical and electrical parts of hydropower projects, which are likely to need replacement (because of mechanical wear or technical obsolescence) every 15–25 years, although the civil engineering elements will last much longer. The neatest way to analyse such a project is to include such costs, called *replacement costs*, in the discounting computations and to let the analysis period end just before another batch of replacement costs is due. For instance, a hydropower plant costs $100 million over years 1 to 5 for initial construction, of which $20 million is estimated to be for mechanical and electrical elements, mainly in year 5. Operation begins in year 6 and it is estimated that these elements will be entirely replaced every 15 years. The analyst inserts replacement costs of $20 million in years 21 and 36 and terminates the analysis at the end of year 50, avoiding counting the next replacement in year 51 (if that cost were included, one would logically need also to include most or all of the years 52 to 65 when the new equipment delivers its benefits). Alternatively, the analysis could be terminated with year 35.

3.5   Inflation, depreciation and loans

The arithmetic of discounting is similar in some ways to that of compound interest and of calculations related to inflation, which can give rise to confusion. The terminology used in some books, and especially in the manuals of computer software, tends to encourage such confusion. For CBA the principle is to express all costs and benefits in a common unit, and this means among other things that all money values used, whether directly in a financial analysis or as a starting point for shadow pricing in an economic or social one, must be expressed in the same monetary unit. In the presence of inflation, a currency unit by itself is not a fixed financial unit: a pound in 2019 is not the same unit as a pound in 2023; the former is slightly bigger. To avoid problems the analyst should use the financial price levels pertaining to a defined date, which should be stated: the unit should be described as 'dollars at mid-2013 prices' or 'rupees at the prices of 1 January 2012'. All costs and benefits should be expressed in that unit, whatever the date they are expected to occur. This unit is labelled at '*constant prices*' or in '*real terms*'. The expected nominal costs in different years, which are sometimes calculated for budgeting purposes, are by contrast referred to as being in '*current prices*', i.e. the prices current in each future year.[6] The defined date itself can be fixed according to the analyst's convenience, for instance the date on which most of the cost estimating was finalised, and is quite independent of 'year 0' for discounting.

It sometimes happens that a particular class of goods is expected to increase in financial price or economic value, or both, in real terms (constant prices). For instance, the real price and value of energy are often forecast to increase in the long term because of depletion of limited resources. If this sort of real future price increase is part of the forecast for a project, it should be included in a CBA. Working as usual in constant prices, the particular item concerned is assumed to increase in price, for instance by a steady 2% per year. This is often called *escalation*, and can apply to either a financial or an economic analysis. (Escalation must not be confused with inflation; measured in current prices the price of such an item would go up faster than the general rate of inflation.)

Adding together a series of numbers representing costs in different years, expressed in current prices, is a misleading exercise comparable to adding together weights expressed in different units such as pounds and kilogrammes, or long and short tons: numbers expressed in different units should, in principle, not be added or compared. Nevertheless, financial operations such as the control of budgets do sometimes require such addition. This gives rise to the term *price contingencies*, which is explained in Box 3.4. It has occasionally been suggested that CBA can be done in current prices and the inflation rate taken into account by fixing the discount rate: this may be useful in financial analysis when loans are involved, but otherwise should be avoided.[7]

Price contingencies are evidently an accounting matter rather than an economic one. Another time-related accounting procedure that sometimes confuses people is *depreciation*, or writing-off. As mentioned in Section 2.2.6, this is a procedure for taking account of the wearing out or ageing of equipment or infrastructure: its nominal asset value is reduced from year to year according to some agreed formula. Even in a financial analysis, however, the annual amounts called 'depreciation' are not cash flows into or out of the relevant party's store of money, so they must not be counted in a CBA. In an economic or social analysis they are similarly irrelevant, being mere adjustments to accounts and not the use or production of any resources or benefits.

Finally, *interest payments* and *loan repayments* can cause doubt. The former are arithmetically similar to discounting, and real interest rates

Box 3.4 *Price contingencies*

Price contingencies is the name given to a numerical difference between sums expressed in constant prices and the same sums expressed in current prices. Suppose, for example, the cost of constructing a project is £40 000 at constant 2018 prices, spread equally over the four years 2018 to 2021, and that inflation is 5% per year over that period. In 2018 the amount spent will be £10 000. In 2019, however, prices will have risen by 5%, and the same amount of work, although correctly valued at £10 000 in 2018 prices, will cost £10 500 because the pounds are now effectively smaller. The third year's work will cost £11 025, and by the year 2021 the pound will have become so much less valuable that the same amount of work will cost £11 576 at 2021 prices.

From a technical point of view, it is meaningless to add up the four numbers because they are attached to four different units (a 2021 pound is not the same unit as a 2019 pound). For financial or banking purposes, however, they do sometimes have to be added. In current terms, with each year's work valued at that year's prices, the cost of the project will appear to have been £43 101.

The differences between constant and current prices, namely 500, 1025 and 1576, are called price contingencies; and the total price contingency is £3101. This computation is sometimes required in advance of a project's implementation, for budgeting purposes and using forecasts of inflation, but it has nothing to do with a CBA and must be clearly distinguished and labelled. Price contingencies must also be distinguished from physical contingencies, which are usually expressed in constant terms, as explained in Section 2.2.

provide one of the guides to an appropriate discount rate in Section 3.9, but their inclusion in a CBA must be decided on the clear criterion of the definition of the party from whose point of view the analysis is done, as discussed in Sections 1.3 and 2.2. In a financial analysis, loans and their repayments and interest payments are real and proper elements of the cost stream, and should be counted when they occur. Suppose, for instance, that an agency borrows all the money to pay for a project that is constructed in years 1 to 3, and pays off the loan, by means of both repayment sums and interest, from year 10 to year 30.

In a normal financial analysis the cash flow will reflect faithfully what money goes into and out of the agency's accounts, although using constant prices. In years 1 to 3 there will be inflows from the lender, and outflows of similar magnitude and timing to the contractors. In years 4 to 9 there will be only the operating takings and expenditures, but from year 10 onwards the outgoings will also include all the payments to the lender. Thus, in effect, the construction costs appear in years 10 to 30. (The repayments are numerically greater than what was paid to the contractors because of the interest, but if the discount rate is similar to the real interest rate the discounting back to year 0 will counter this apparent increase in cost.) In an economic analysis of the same project, however, the financial transactions between agency and lender will be ignored and the costs will appear, shadow priced if appropriate, when they are physically incurred, that is in years 1 to 3.

There is a third kind of analysis, namely a 'financial analysis before financing'. This counts the money inflows and outflows of a project's implementing agency as if it were managing without loans; in this example the costs would count when the contractors were paid. Such an analysis helps to identify what sort of loan package a project needs. It is advocated by the World Bank handbook (Belli *et al.*, 2001, p. 28).

3.6   Choice of indicator for particular purposes

Which indicator is used depends mainly on the nature of the decision that is to be guided by the analysis, although it is also in some cases a matter of the decision-maker's preference, or of habit, or convention in a particular sector, region or context.

If the analysis is to be used to decide whether or not to implement a certain project, an indicator is normally compared with a predetermined threshold value that determines whether a course of action is acceptable or not. This is normally a *test discount rate*. For instance, a group of decision-makers, such as a government department or a development bank, may determine in advance that all projects showing positive NPV in economic CBA, discounting at 11%, will be regarded as economically acceptable. It is evident from the arithmetic of discounting that any project with a positive NPV at 11% also has a B/C ratio at 11% that is more than unity, and also has an IRR above 11%. So there are three ways of stating the same acceptability criterion: NPV at 11% must be positive, B/C at 11% must be more than one, or IRR must be over 11%. For this kind of analysis it matters little which indicator is used, as the result will be the same with any of the three. The critical value can be seen either as an appropriate and agreed test discount rate or as a limiting IRR. What value is appropriate is a potentially controversial matter, and is the subject of Section 3.9.

To choose between projects or courses of action that are *mutually exclusive*, in the sense used in Section 1.3 (for technical reasons, not just because of capital shortage), the criterion should be maximum NPV with an appropriate discount rate. Maximising any other indicator may lead to a wrong decision, especially when the competing projects vary widely in size. Box 3.5 gives a demonstration of this, using the concept of a differential project. This demonstration can be used to convince any CBA users who have doubts about the correctness of the use of NPV for the mutually exclusive case.[8]

A special case of the technically mutually exclusive type of analysis is a *least-cost analysis*, when benefits of both or all alternatives are the same and do not need to be valued. The cost that is minimised is, of course, the NPV of costs. Some people compute an *equalising discount rate*, similar to an IRR, namely the discount rate that would make two alternatives have the same NPV. As a decision rule this boils down to the same as maximising NPV at the test discount rate. An example is given in Chapter 5.

The next type of decision is the *ranking* problem: there are a number of projects competing for a limited stock of available resources (in financial terms, for a limited budget). The limited resource may be a fixed quantity irrespective of time, or a fixed set of annual budgets in the future. In either case, planners need an indication of the order of priority in which projects should be implemented. To rank by NPV would be no help, as a large project that was barely acceptable would be given priority over a small one with excellent benefits relative to its small costs. The correct procedure is to rank the projects in descending order of B/C ratios. It can be shown that ranking by IRR can lead to bad decisions, although in practice the ranked order using IRR would often be the same as the correct one using B/C. Among the various kinds of B/C ratio discussed in Section 3.3, the net benefit/investment ratio (N/K) is, strictly speaking, the correct one to use for a single period, if investment funds are the limiting resource.[9] The more general principle is to look for the best possible return (in a financial, economic or social sense, as appropriate) to each unit or increment of whatever resource is thought to be the limiting one.

With all kinds of CBA, each alternative or project should, in principle, be *internally optimised* and *optimally packaged* before its indicator

Box 3.5 Choice of indicator for the technically mutually exclusive case

Suppose a CBA is to guide the choice between a small and a large dam, at one and the same site. The larger dam costs more but gives greater benefits. After consistent analysis of both these competing projects, using an agreed test discount rate, it is found that the discounted economic costs are 40 units (e.g. million rupees at 2017 prices) for the small dam and 80 units for the large one. The discounted benefits are 50 and 96 units. So the NPVs are 10 units for the small dam and 16 units for the large one (50 − 40 and 96 − 80), which seems to favour the larger dam. But the B/C ratios are 1.25 for the small dam and 1.20 for the large one (50/40 and 96/80), which would seem to favour the smaller dam. The IRR would also be higher for the smaller dam.

To show why the larger dam is better and NPV is the right criterion, consider a hypothetical differential project called 'large dam instead of small dam'. If it has been decided, at least, to spend the 40 units of resources and to build, at least, the small dam, this differential analysis will show whether it is economically acceptable to spend the extra 40 units and build the large dam instead. Evidently this differential project has a discounted economic cost of 40 units (80−40) and a benefit of 46 units (96−50), and thus an NPV of 6 units and a B/C ratio of 1.15. Although these indicators are lower than those of either dam, the NPV is still positive and the B/C ratio is more than one, so the IRR must be above the test rate and the differential project is acceptable.

If there were several competing dam sizes to be analysed (and capital resources were not a relevant constraint), it would evidently be economically desirable to choose the one giving the greatest NPV: the differential project between this and the next smaller one would be acceptable, but the differential project between it and the next bigger one would have a negative NPV, a B/C ratio less than one and an IRR below the test rate, so it would not be acceptable. To choose any other dam size, such as, for instance, the one with the highest B/C ratio or the one with the highest IRR, would leave an economically favourable enterprise un-implemented. An example is given in Chapter 5.

is compared with something else or with a test value. 'Internally opti-mised' means that all significant lower-level (design-type) decisions should have been rationally made, using CBA where appropriate. 'Optimally packaged' means that, if a project is a composite of several components, of which some could be omitted (or varied in size or timing), then the composite project should be the best possible such package before it is compared with other projects. Both conditions can be met by formulating various versions or variants of the project and comparing them against each other as technically mutually exclu-sive options. This comparison or optimisation within the one project should be done in a way consistent with the later comparisons against other projects or test values. This means that, ideally, the lower-level optimisations should use the same shadow prices and discount rate as the intended final analysis. In practice one can seldom meet these conditions totally and strictly, but the aim should be kept in mind, and any compromise that may be necessary (e.g. to get the analysis done in a reasonable time frame) should be made consciously and should preferably be mentioned in the report.[10]

To summarise, the rules-of-thumb for choosing which indicator to use for a given CBA are:

- For a yes/no decision on one project's implementation: use NPV, B/C or IRR, with a previously agreed test discount rate or threshold IRR.
- For optimisation between technically mutually exclusive alterna-tives: maximise NPV.
- For ranking of mutually independent projects that compete for limited resources: use a B/C ratio, preferably N/K.

The choice of indicator can be controversial. Some decision-makers are accustomed to using indicators in ways that are not correct, especially ranking by IRR. Sometimes the analyst needs to argue for a particular procedure, giving examples to show the consequences of a choice.[11] It can be seen from this summary that one can cover all cases without using IRR at all, and some decision-makers (notably the British Treasury) will be happy with this. Its use is, however, so firmly established in the procedures and habits of some people and institutions that analysts often need to calculate and report it, if only as an extra alongside what they regard as the main indicator. IRR is historically and logically linked to the financial concepts of return to investment and interest on a loan, and tends to be the most familiar indicator to bankers, including development bankers (although some are accustomed to using N/K or some similar ratio). Some people argue that using the IRR saves everyone the trouble of agreeing a test discount rate. This is wrong: for yes/no decisions one needs a test rate anyway, to compare the IRR against, and for optimising and ranking decisions the use of IRR is incorrect. Some cases where this section recommends the use of NPV can also be handled using an equalising discount rate between two alternatives, which amounts to the same thing but is less transparent. People with a special affinity for IRRs sometimes prefer such a procedure.

3.7   Sensitivity tests

The parameters that quantify costs, benefits and time effects in a CBA are always in some degree uncertain. In some cases, such as engineering cost estimates, it would require a great deal of time and energy to increase the precision of the numbers. For other parameters, such as the discount rate and many shadow price factors, value judgements are involved and there is, in principle, no knowable correct number. In either case it is useful to find out how sensitive an analysis' result is to variations or errors in the data and assumptions. This can some-times be done analytically, but the most transparent, flexible and general way is by means of *sensitivity tests*. Once an analysis has been carried through to the point of calculating the appropriate indicator, using the best available estimate for each parameter, the parameter values are arbitrarily varied one by one and the computation is repeated each time to find out how the indicator varies with each input para-meter. If the analysis is done on a computer, for instance in a spreadsheet, most such tests only take a minute or two each.

Taking as an example the hypothetical project considered in Boxes 3.1 and 3.2 and Table 3.2, the first step is to define the analysis with the best estimates of all parameters as the *base case* (this case has already been shown in Table 3.2). The next step is to choose one parameter, for example the benefits. A simple way of varying the benefits (but not the only way) is to consider the possibility that, whether because conditions change or because the estimates were overoptimistic, the benefits will all turn out to be 10% less than estimated in the base case. Recalculating the example with everything the same except that the benefits are all 90% of what they were in the base case (i.e. from year 4 they are 9, 27, 81, 144, and then always 180 units), we find that the MPV becomes 354 units instead of 477 units and the B/C ratio becomes 1.47 instead of 1.63. Reducing the benefits still further, to 80% of the base case ones, gives indicator values of 231 and 1.31. Raising the benefits by 10%

gives indicator values of 600 and 1.80. These numbers show that the change of indicator value per unit change of input parameter is fairly uniform and symmetrical about the base case: the NPV changes by 12.3 units, or 2.6% of its base case value of 477 units, for each 1% change in benefit levels, while the B/C ratio (naturally) changes by exactly 1% for every 1% change in benefits. This set of sensitivity tests, along with others to be discussed, is shown graphically in Figure 3.1.

A similar set of computations can be done on other parameters. If the investment costs (those in years 1 to 4 in this example) are increased by 20% the NPV goes down by 141 units, from 477 units in the base case to 336 units, and the B/C ratio goes down from 1.63 to 1.38. Again, sensitivity tests with other percentage changes to investment costs show that the change is roughly uniform and symmetrical, with NPV changing by 1.5% and B/C by about 0.9% for every 1% change in the investment costs. These ratios of percentage indicator change to percentage change are sometimes called *sensitivity indicators*.

These sensitivity tests are introductory examples, easy to understand and quite useful for giving decision-makers a feel for the importance of different parameters, but are not especially informative because these sensitivities are fairly obvious anyway. (As the benefits provide the whole of the numerator of the B/C ratio, it is no surprise that a 1% change in benefits gives a 1% change in ratio: it is a statement of the obvious. The fact that a 1% change in investment costs gives a 0.9% change in B/C ratio merely reflects the fact that investment costs contribute most, but not all, of the cost stream.) A sensitivity test that is more useful, because its result is less predictable, is one on the recurrent cost from year 5 onwards. The base case in Table 3.2 has recurrent costs of 7 units/year: if this is reduced to 6 units/year the NPV only changes marginally from 477 to 484, or about 0.1% per 1% change in recurrent cost. Even raising the recurrent cost to 10 units/year instead of 7 units/year only shifts the NPV to 455, again a tenth of the relative change in the input parameter. The B/C ratio is similarly insensitive to recurrent cost. This means that the result of the CBA, in terms of its indicator values, is not drastically affected by uncertainty in the estimate of recurrent cost (because that forms only a small part of total discounted costs, being swamped by the investment component). This is a useful fact because it tells all concerned that it is not worthwhile to put much effort into refining or arguing about that particular estimate; effort should instead be concentrated on parameters that show high sensitivity.

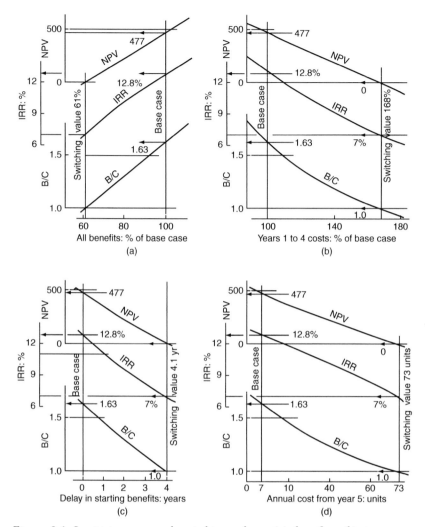

Figure 3.1 Sensitivity tests and switching values: (a) benefits; (b) investment costs; (c) benefit timing; (d) recurrent costs

Another estimate that might be uncertain or debatable in this analysis is the timing of the benefit start-up. A simple sensitivity test might be done on this by assuming that the whole benefit stream starts a year later than the base case schedule: 10 units in year 5 and reaching the steady 200 units/year only in year 9. Making this change in the discounting computation produces a reduction in NPV from the original 477 units to only 348 units, and the B/C ratio comes down from 1.63 to 1.46. These are quite drastic reductions, so it can

63

be stated in a report or presentation to decision-makers that 'the result is quite sensitive to the timing of benefit start-up'. Shifting the start-up by two years almost doubles the effect.

A further statistic that is occasionally used is a *switching value* of a particular input parameter. This is the value of that parameter which, if substituted for the parameter's base case value, will bring the indicators to their critical or threshold values. If the above example is a yes/no analysis, the threshold values are zero for the NPV, 1.0 for the B/C ratio, and 7% for the IRR. By trial and error in the spreadsheet version of Table 3.2, it can quickly be found that the switching value for investment costs is 168%: these costs would have to rise 68% above the base case values to bring the three indicators to their threshold values, thus almost reversing (switching) the indicated decision from yes to no. The switching value for recurrent costs is 73 units instead of 7 units, which should be far outside any conceivable margin of bad estimating and reflects the low sensitivity of this parameter. The switching value for benefit delays is just over four years, which is not unthinkable: a five-year delay in getting benefits flowing (which can easily happen if contractors fail or war interrupts a project) would turn a strongly favourable economic result into an unacceptable ('uneconomic') project.[12] (Switching values for delay should be used with caution, as the effects of delay may be non-linear and dependent on the cause of the delay.)

In practice, an analyst should usually do several sensitivity tests and present in a report only a summary of the results, such as the words just quoted concerning benefit timing, backed up by a few numbers. This whole example could be summed up by saying that

> *the result is quite sensitive to the timing of benefit start-up, moderately sensitive to the general levels of benefits and of investment costs, and fairly insensitive to recurrent cost estimates.*

All these tests can be carried out with the IRR as well as with the two indicators discussed here, showing similar sensitivities.

Figure 3.1 shows these sensitivity tests graphically, including IRR as well as the other two indicators. Those readers who prefer graphs to numbers will see that the sensitivities are represented by the slopes of curves, although these can only be interpreted if account is taken of the scales used for plotting. The graphs also show how smooth the curves are, most of them being nearly or completely linear, which is typical for most projects with normally shaped cost and benefit streams. A consequence of this is that it is usually redundant to quote indicator changes for 10% parameter changes when those for 20% have already been given (they will always be roughly half), or to give results on both sides of the base case. In practice, it is usually sufficient in reporting an analysis to give the effect on key indicators of a 20% change in main input parameters, this being the sort of mis-estimate or change of circumstances that a decision-maker can easily picture. Where possible, a parameter change related to a physically significant difference is better than an arbitrary percentage. For a civil engineering project one test could examine the consequences for CBA results of discovering a lens of soft clay at a 6 m depth in the foundations, this being considered unlikely but possible in the light of the site investigations; another could show what would happen if a large flood overtopped a cofferdam in year 3, as at Kariba on the Zambezi, even though this is very improbable. As demonstrated by one of the examples presented in Chapter 5, the change tested is usually in the direction that would tend to reverse the decision indicated by the base case.

The parameters tested should usually include estimated costs (preferably several specific subcategories such as recurrent costs or foundation works), benefits (particularly any tariffs or unit rates used in the analysis), demand growth forecasts, and the principal shadow pricing factors or related assumptions.

In all this discussion so far we have changed only one parameter at a time, leaving all the others at their base case values. This is the recommended way to conduct and report sensitivity tests, except when two or more of the parameters are correlated, as the effect on an indicator of changing two or more independent parameters simultaneously, although easy to compute, is difficult to interpret. (When two parameters, such as delay and cost increase, are correlated because they are affected by a common cause, it may be best to formulate a sensitivity test related explicitly to that common cause, so the changes in the two parameters become mere intermediate variables in that test.) The exception may be right at the end of a CBA report, when the robustness of a result can be emphasised by saying something like:

the result is favourable even if costs increase by 20% and benefits decrease by 10% and start a year late.

It is not unknown for an analyst to go a step further in this direction and present a graph called the *locus of switching values*, which shows what combinations of variations in two parameters would make indicators critical: for instance +65% on costs, or −40% on benefits, or +30% on costs together with −20% on benefits, or +95% on costs with

+15% on benefits. This may be a convenient communication tool for some report readers, but adds little information that is not contained in the separate statements about the two parameters.

Further remarks on presentation of results are given in Chapter 5: the graphs in Figure 3.1 would not be suitable for a CBA report, as they give far too much information and most of it is repetitive.[13]

Sensitivity tests are ideal for investigating and reporting on uncertainty, in the sense of variations away from the best-estimate due to the inherent difficulty of estimating and forecasting with limited information (see the discussion of physical contingencies in Section 2.2). The distinction between *risk* and *uncertainty* is not clear-cut, and the more useful distinction is that between non-statistical uncertainty about unprecedented events, such as the future traffic on the Dunsinane Bypass, and statistical uncertainty about repeatable events, such as floods. (I cannot guess whether a particular bridge will suffer a certain sort of very large flood in year 17, but I can, through standard hydrological analysis, make an estimate of the probability of that event, along with the probability of any sort of flood in any year.) Appendix A discusses uncertainty in some detail, including the best ways to use sensitivity tests.

Another way of investigating and reporting uncertainty is to run the analysis many times with varying values of the main input parameters such as costs or timing. This probabilistic analysis is often called *risk analysis*, although that term has other uses also, and the term 'Monte-Carlo analysis' may also be used because of the sampling techniques involved. This is increasingly favoured by funding agencies such as the World Bank or national aid agencies.[14] It is discussed in Appendix A.

3.8   Terminology and reporting

Section 1.6 has argued for the transparent presentation of an analysis by a candid and honest analyst to a well-informed decision-maker, and Chapter 5 discusses ways of presenting a CBA. The core of the reporting is the presentation of indicators. Good reporting requires that, before the indicators are given, clear statements should be made of the precise way in which benefits and costs have been defined and valued. This is especially necessary if any kind of B/C ratio is being reported. As discussed in Section 3.3, some project effects can reasonably be regarded either as positive benefits or as negative costs. Which of these is chosen makes no difference to an NPV or an IRR, but does affect a B/C ratio.

When it comes to reporting the indicator values, analysts often finish reporting an economic CBA by saying that a project whose indicator has reached or exceeded some predetermined threshold value is economically *feasible* or *viable*. This is potentially misleading, as these words normally mean able to be done, achievable, attainable, practicable or workable. It may be meaningful to say that a project is *financially* not feasible or not viable, meaning it would not survive for financial reasons, but an economic CBA refers to whether a project is desirable or acceptable according to some defined criteria. A project that is economically or socially undesirable may be quite feasible and viable, and many would argue that such projects do in fact get implemented and carried through to completion. Conversely, as demonstrated by one of the examples in Chapter 5, a project can be found to be economically justifiable but financially not viable. Financial viability may be an overriding decision guide alongside economic CBA, and such a project will probably not be implemented. The World Bank handbook puts it mildly:

if the project's financial performance is in jeopardy, its economic performance may suffer.[15]

Nevertheless, for better or worse, the words 'feasible' and 'viable' are often used in the CBA context to indicate economic or social acceptability. The reader is advised to avoid using them. A similar, but vague, term is used to describe a project as 'economic' or 'uneconomic'. If this is merely a convenient shorthand for economically acceptable or desirable, as in Box 2.3, page 20, then it is defensible, but it runs the risk of appearing to imply an inherently correct technical measurement rather than the result of an analysis founded on many value judgements. A much clearer and better differentiated terminology is found in the following quote from the terms of reference for a planning study:

Phase 2 will proceed only if the Phase 1 report shows that the project is technically feasible, economically justifiable, and financially viable.

The word *sustainable* is sometimes used in place of 'viable'. Both are broad concepts concerning what will survive and flourish and what will not, and can be used in financial, environmental, social, institutional and political senses. Whether a project fails because of an earthquake, or the indifference of the intended beneficiaries, or inter-ministry bickering in the government, or lack of maintenance, or bad design, the fact remains that it was not viable and sustainable.

3.9   The appropriate discount rate

At the beginning of this chapter the concept of discounting was introduced by starting from an individual's time preference rate. The further assumption of a collective time preference rate for a whole society or nation was made without discussion or justification. Although time preference feels natural to most people, and is often applied unconsciously, it is quite difficult to explain and justify why it should be applied systematically to collective decisions, and more difficult to set out a coherent and convincing justification for any particular discount rate. Among specialists in the field there are many approaches and shades of emphasis, and the subject is continually debated in the professional literature. The reader of this book will not normally need to know the details, but should have some acquaintance with the theory and the different approaches, so as to be able to understand and contribute to discussions of different discount rates as well as to explain to decision-makers the basis of an analysis. This section presents a brief summary of the matter, and Appendix E gives a more detailed account, together with some references to the extensive published literature.

The discount rate discussed here refers to computations done in constant-value terms (Section 3.5), and has nothing to do with inflation. Because it relates to people's decisions about when they will use resources and how long they will wait for benefits, the test or threshold discount rate used to guide decisions is, in economic terms, connected to the balance between consumption, on the one hand, and savings and investment, on the other. This is because, if CBA is used consistently to decide which projects are implemented, a high test discount rate will tend to reject more investment projects than a low one, and so will lead to lower levels of investment and a higher proportion of resources and income being spent on consumption.

The choice of test discount rate also concerns equity between generations, which is an ethical and political matter rather than a technical one. If my children are about 30 years younger than me and I use or approve a discount rate of 11% per year, I am implying that their consumption when they reach my present age is less than one-twentieth as significant, for my decisions, as my consumption now (the relevant PV/FV factor in Appendix J is 0.044). If, however, I use a discount rate of only 5%, their consumption is treated as worth almost a quarter of mine (a factor of 0.23).

Even for a simple financial analysis there is no such thing as a single, unambiguous, correct discount rate, and for an economic or social analysis the choice, with its ethical implications, is ultimately a political one. The various ways of thinking about the discount rate fall into three broad groups:

- time preference as an observable human characteristic
- the opportunity cost of resources, especially of capital
- the capital rationing device.

It is observed that individuals prefer good things to happen sooner rather than later, and bad things later rather than sooner. By implication they apply a positive discount rate to their individual decisions, and economists can observe people's behaviour and deduce what their individual time preference rate, or unconscious discount rate, is. When decisions are made on behalf of a large number of people, such as a community or society, there is again an observable time preference rate: this collective one is called the *social time preference rate*. However, this rate is generally lower than the time preference rate the people apply as individuals acting in a market situation, because they are aware of desirable things that no market can provide and are conscious, however vaguely, of the ethical rights of future generations. It is widely agreed that those who act for people at the collective level, i.e. governments, have a responsibility firstly to reflect this difference between individual rates and what individuals wish for collective rates, and perhaps to go beyond that and represent the interests of future generations. Because this implies taking a long-term view, the social time preference rate is probably quite low, for instance in the range 2–6%.

The second approach is to apply the opportunity cost principle to discounting. The discount rate should reflect what society forgoes elsewhere if it uses certain resources for a period of time in a certain project. It foregoes whatever other use might have been made of the resources, or whatever investment might have been made with the money. What that foregoing costs will, of course, depend both on the amount of resources and the length of time for which they are tied up in the project. When money is borrowed by a developer, the interest he pays to the lender is the borrower's cost and the lender's reward, primarily a reward for waiting. Strictly speaking, the relevant number is the real interest rate or return on the last or marginal investment, i.e. the one that uses up the last of the available capital. This number, which is often used as a guide to the appropriate discount rate, is called the *social opportunity cost* or the *opportunity cost of capital*. The actual determination of the opportunity cost of capital is a job for professional economists; typical rates are in the range 8–15%, with 10–12% very commonly being used. In principle, it can be expected always to be

> ## Box 3.6 Real and nominal interest rates
>
> Consider a country with a steady inflation rate of 5% per year, in which a certain bank offers a nominal interest rate on deposits of 15.5% per year. If, in 2019, I invest a hundred 2019-dollars with this bank, I will expect after two years to receive back \$13 340 (1.155^2 is 1.3340). But they will be 2021-dollars. The buying power of a 2021 dollar will be less than that of a 2019-dollar because of two years' inflation at 5% per year, i.e. by a ratio of 1/1.1025 (1.05^2). So the 133.40 2021-dollars that I receive will be worth only 121.00 2019-dollars (133.40/1.1025). Measured in consistent units, therefore, my investment has increased from 100 to 121 units in two years, an annual increase of 10% (1.10^2 is 1.21). Calculating in constant units or in real terms shows the real interest rate to be 10%, in contrast to the nominal interest rate of 15.5%.
>
> The true arithmetical relationship between the two interest rates and the inflation rate is thus (1 + real rate) = (1 + nominal rate) ÷ (1 + inflation), the rates being fractions rather than percentages. In this example 1.10 = 1.155 ÷ 1.05. When the rates are fairly small, it is a common approximation merely to subtract the inflation rate from the nominal interest rate, which in this case would give the erroneous result of 10.5% for the real rate.

higher than the real market rate of return on investments, but the starting point for arriving at a discount rate under this approach is normally an examination of the interest rates prevailing for certain types of investment. (That analysis must also consider risk and management costs.) In the presence of inflation it must be the real interest rate that is used, because the CBA will be done in real terms with constant prices. Box 3.6 explains the relationship between real interest rate, nominal interest rate and inflation.

Many argue that the government, as the representative of collective interests, should undertake investments that the private sector does not find attractive. This implies that collective decisions (involving economic and social CBAs in the sense of Section 1.3) should be guided by a lower discount rate than private ones.

The third way of looking at the discount rate is as a *capital rationing* device in a financial sense, or a resource rationing device in a more general sense. If a particular country or sector has a limited amount of capital or other resources, and those resources are allocated to projects by judging each project's CBA indicators against a test discount rate, then the higher the test rate the fewer projects will be approved and implemented. If the test rate is too high there will not be enough projects being implemented to use up the resources, and if it is too low the resources will run out. In principle, there is an ideal rate somewhere in between that will just achieve the full use of resources; this is the appropriate discount rate for a capital rationing role. In contrast to the other two approaches, which try to work out what the discount rate ought to be according to some theory or set of ideals, this is a pragmatic approach that merely accepts what the discount rate actually does in practice (i.e. regulate the stream of projects that get implemented) and looks for a rate that works. The discount rate arrived at by this approach is only as right or appropriate as the supply of resources, and that may be far from optimal.

As indicated above, the social time preference approach will generally indicate a relatively low test discount rate, and the opportunity cost of capital approach a high one. The capital rationing approach may do either, but it tends to be used in places where resources are scarce, as an argument for a high test rate, and to be ignored in places where it might indicate a low one. Specialists, particularly economists, will often have their own view about the appropriate test discount rate, and will in some cases undertake special studies to ascertain a figure, but most readers of this book will normally arrive at a rate for each analysis by asking someone else, typically the main decision-maker or the party that is likely to finance the project. It may be valuable for the analyst to discuss the choice of discount rate with the decision-makers, explaining the various approaches if they are not already well understood. Sometimes a sensitivity test is worthwhile, to demonstrate how the indicated decision depends on the discount rate. Governments often have planning commissions or similar agencies who oversee CBAs and other studies in many sectors, and such an agency will usually be able to state an approved discount rate for any analysis, possibly bearing in mind regional and intersector factors.

Interest rates usually have, alongside their reward-for-waiting role, an element of reward for risk-taking, and this is sometimes explicitly labelled as a *risk premium*. To some extent this concept is carried over into the choice of discount rate for CBA, if the opportunity cost of capital is to take some account of the riskiness of a project. It is, however, clearer and more transparent to keep the whole matter of risk and uncertainty separate from matters of timing, dealing with it explicitly, preferably by means of sensitivity tests and perhaps probabilistic analysis (see Appendix A).

Although in economic theory the efficient allocation of resources requires that the same rate be used in all sectors of a national economy, there tend in practice to be different rates for different purposes or sectors, and for different categories of decision-maker. If the context is dominated by bankers or financiers, the rationale of the opportunity cost of capital tends to predominate, sometimes accompanied more or less explicitly by capital rationing considerations, and the rate used is usually in the range 8–15%, typically 10–12%. (High rates are sometimes used to counter the suspected optimism of planners, although it would be better to improve the estimating.) If the context is more one of political decisions taken on behalf of a nation, with a long-term view and ethical considerations, the collective (social) time-preference-rate approach is more likely, and the discount rate may be as low as 3%. Another distinction that is occasionally relevant is that the opportunity cost of capital is related to production or producers, and the social time preference rate to consumption or consumers. Some people advocate a conscious weighted-average discount rate that takes account of all approaches. Whether it is stated or not, the rates used in practice do often originate from some compromise between two or more approaches. In the more developed economies the approved rates for public sector decisions often lie in the range 3–10%, while in less developed countries the rate is typically at least 10%. There is, however, a tendency to vary the rate according to the sector involved, as a conscious political decision to favour some sectors more than the application of a standard rate would. Decisions with environmental implications are often guided by CBA using a relatively low discount rate (or some special discounting method that gives a similar effect), because of the perception that long-term consequences are important. The British Government's guidelines for public sector decisions specify a normal rate of 3.5%, with lower rates for discounting after year 30.[16]

Example (for readers to work through)

This example illustrates both the arithmetic and the interpretation of discounting, and helps to give the reader a feel for the effect of discounting at different rates and times. For the arithmetical parts there are unique correct answers, and these can be found in Appendix H. For the discussion parts there are no such unique answers, but model answers are provided in the same appendix. Readers who do not wish to spend the time to work through all the questions can still benefit by reading them and thinking about how they would be dealt with.

The arithmetic can be done either by using the tables in Appendix J or by means of a computer spreadsheet, for which the guidance in Box 3.3, plus the reference manual or help function of the particular spreadsheet software being used, should be sufficient. Readers using a spreadsheet can take the opportunity to design it to serve not only to perform the calculations but also as part of a report (intermediate calculations that are not of interest to a report reader can be placed off to one side or on another sheet, outside the range that will be printed). To make sensitivity tests easier, it is wise to make significant parameters such as discount rate and recurrent costs and benefits into spreadsheet variables rather than building them into formulae.

(a) The large proposed project Trillian 5 is the subject of a feasibility study for the Government of Utopia, which hopes to obtain loan funds later from the Galactic Bank. You are a senior employee of the consulting firm Arthur Dent Associates (ADA), engaged on the study and responsible for the cost–benefit analysis. With whom would you discuss the choice of discount rate?

(b) It has been agreed that the discount rate for the base case analysis is to be 7% per year, and the best-estimates of the project's costs and benefits, in millions of Utopian dollars (MUt\$) at June 2003 economic prices, are as set out in Table 3.2. Check the arithmetic of that table. Find the economic internal rate of return (EIRR). (If you are using a computer, the EIRR can be built into the spreadsheet very easily. If not, it requires tedious computations and trial and error or graphical interpolation, so readers using tables may prefer to skip the calculation of the EIRR: the rest of this exercise does not depend on it.)

(c) Just before the report is finalised, the engineers find an error in their estimates and change the annual recurrent benefit from year 8 onwards, from MUt\$200 to MUt\$180. In a hastily convened meeting that verges on panic until you explain how little this matters, the directors of ADA decide that, as the report absolutely must be finished by Friday, the estimates for the build-up period in years 4 to 7 can remain as before. The directors ask you what the change is likely to do to the economic indicators; they do not want to wait more than one minute for your answer. Without reworking any discounting calculations, make a quick estimate of

73

the effect on B/C and NPV, and comment to the directors on the implications.

(d) Now rework the full calculation with the new estimate of recurrent benefits. Do you need to revise what you said to the directors?

From now on the base case is like the one in Table 3.2 but with MUt$180 instead of MUt$200 for annual benefits from year 8.

(e) Confidence in the estimating abilities of the firm's engineers has been shaken a little, and people are wondering what would happen if all the benefits were significantly lower than they say. Design and conduct a sensitivity test to investigate this. Draft a paragraph for the feasibility report, describing and interpreting this test.

(f) Seismologists say that, although the project in its final form is designed to resist significant earthquake damage, an earthquake near the end of year 2, which is unlikely but possible, would severely interrupt construction. All subsequent costs and benefits would be delayed by a year, and the whole of year 3 would be taken up with repairing the damage at an economic cost of MUt$40. Estimate the effect all this would have on the economic indicators. Roughly how much of this effect is due to the delay, and how much to the repair cost? Draft a section for the report, avoiding alarming the decision-makers unduly, but ensuring they are well informed about this danger. Think about what you would say to a judicial enquiry after the earthquake actually happened.

(g) Rework the base case analysis with a longer analysis period, ending in year 43 instead of year 20. Comment on the differences.

(h) Suppose now that you are analysing another project, Seizewell E, whose base case analysis is, by an amazing and convenient coincidence, exactly the same as that of Trillian 5 up to year 20 (with annual benefits of MUt$180), but which requires replacement of turbines at 20-year intervals. They were originally installed in year 4 and each replacement has an economic cost of MUt$42 (still using June 2003 prices). Annual recurrent costs continue after year 20 as before it, and with this attention to maintenance, plus the regular replacement of turbines, the project can continue indefinitely. What analysis period would you use for the CBA? Run the base case analysis with periods ending in year 23, year 43 and year 63. Do the differences matter?

(i) You discover by chance that the engineers had forgotten (or cynically omitted) to tell you that Seizewell E is a nuclear power plant: they expect that it will operate until the end of year 43 (Why is that a convenient year to choose?) and then be decommissioned. All benefits will cease at the end of year 43, and the decommissioning will cost MUt$450, evenly spread over years 44 to 46. Thereafter the maintenance of the site, with heavy security because of residual radiation, will cost MUt$3 per year until year 100, but after that the site will be considered safe and there will be no further significant costs. Amend the base case to take account of all this. Which estimate is more significant: the MUt$450 for decommissioning, the annual MUt$3 thereafter, or the cessation of security arrangements in year 100?

(j) With these revelations, Seizewell E is obviously a project with long-term implications, and the choice of discount rate may be significant. What would be the effect on the economic indicators of using rates of 3% and 12% instead of 7%? Comment on the results.

(k) The feasibility report on Seizewell E is published and praised, and you become a director of ADA. The proposed project goes to a public enquiry. On day 42 the tedium is relieved by a member of the public who implausibly gives her name as Ford Focus; she challenges the choice of discount rate and the postoperation-phase estimates in an impassioned speech about the fate of her yet unborn grandchildren who might have to live with the remains of Seizewell E. The representative of the Utopian Department of the Environment, who agreed to the rate and estimates at the time of the study, suffers a convenient asthma attack and leaves the hall. The judge turns to you. How do you respond to Ms Focus' outburst?

Notes

1 Sonnet XVI, William Shakespeare; Thorpe, London, 1609 (ed. W.J. Craig, Oxford University Press, 1905).

2 Some textbooks tabulate up to four other ratios that are merely sums and reciprocals of the two used here, or give them financial-sounding names such as *present value of annuity* for PV/RV and *capital recovery factor* for its reciprocal. For the reader capable of dividing as well as multiplying, the two ratios tabulated in this book are all that are needed.

3 This case is described in many textbooks, for instance Curry and Weiss 1993 (Appendix 3.2, p. 72); see also Eckstein and Lecker (1995).

4 This analysis has already been mentioned, as example (c) in Section 2.1.

5 This, with the abbreviation N/K, is the name used in the authoritative World Bank book (Gittinger, 1982). Others call it the *present value over*

capital ratio, PV/K (Irvin, 1978, Section 1.08) or NB/K (Yaffey, 1994; Yaffey also discusses the other authors). The indicator is also defined by Perkins (1994, Section 5.3).

6 The World Bank handbook uses the term *nominal prices* rather than *current prices* (Belli *et al.*, 2001).

7 A case for doing financial analyses in current prices is made in Yaffey (1992, Ch. 12) and discussed in Potts (1996). Calculation of price contingencies at various levels of precision, and with various numbers of currencies, is discussed in Ward and Deren (1991, Ch. 20). Yaffey (1992) uses the term *price contingency* in a different sense, to mean the part of what is here called *physical contingency* that is due to changes in unit costs rather than changes in physical quantities.

8 A longer explanation is given in the technical appendix in Belli *et al.* (2001); various sorts of technically mutually exclusive decisions are discussed in Gittinger (1982, pp. 373–393). Perkins (1994, Section 5.2.3) shows explicitly why IRR is the wrong indicator for this purpose.

9 For a full exposition see Perkins (1994, Sections 5.1.4 and 5.2.4).

10 This was emphasised in the 1994 version of the British Treasury guideline. The World Bank handbook gives an example of seven different ways of packaging a project with potential hydropower, irrigation and recreation benefits (Belli *et al.*, 2001, pp. 22–23).

11 A discussion of the various indicators from the World Bank viewpoint, particularly for ranking and for mutually exclusive alternatives, can be found in Gittinger (1982, pp. 358–361). The official British recommendation is to use NPV (H.M. Treasury 2003, Chapter 5). The case where IRR gives a bad ranking decision is set out in Irvin (1978, Section 1.08) and discussed in Finney (1990). A purely financial discussion is given in Brealey and Myers (1984, Section 5.1). Yaffey (1994) reviews earlier literature and discusses in detail the effects of particular indices.

12 The World Bank handbook prefers switching values to other ways of conducting and presenting sensitivity tests (Belli *et al.*, 2001, Ch.11).

13 Some people draw attention to the *gross/net ratio* in the recurrent costs and benefits, during the steady operation phase of a project, as an indicator of the level of uncertainty in a CBA. In the above example, the base case annual discounted benefit and costs are 200 units and 7 units, respectively, so the annual net benefit is 193 units and the ratio of gross to net benefit is 200/193 = 1.04, a very small ratio. If a project has a high gross/net ratio, then the annual net benefit is a small difference between a large gross benefit and a large cost (like any small difference between large numbers this is very sensitive to estimating errors in either of the large numbers). This can be the case for the analysis of a rehabilitation project (Yaffey and Tribe, 1992), or for an incremental analysis. However, this sort of sensitivity is fairly obvious, and mathematical demonstration of it is seldom appropriate. A hypothetical examination of the effect is given in Yaffey (1994).

14 See the World Bank handbook (Belli *et al.*, 2001, Ch. 11) and H.M. Treasury (2003, Annex 4).
15 World Bank (1996, Ch. 2).
16 H.M. Treasury (2003, Annex 6). Such fixed or recommended rates can be changed from time to time, and an analyst should always enquire whether a government guideline is still current.

4

Application to particular fields

4.1   General

The first three chapters have presented the principles and basic techniques that make up the toolkit of CBA, and the appendices give some details. The purpose of this chapter is to illustrate those techniques in the context of several different fields, and to introduce the reader to some of the ways that CBA can be adapted to a variety of decision-guiding tasks. The order of the sections does not indicate their relative importance, but is chosen to introduce concepts which in many cases apply to several fields. The use of gross and net returns is introduced in the context of agricultural projects in Section 4.2 but is applicable to other fields also. Cost–utility analysis is described in the health context, while average incremental cost and tariffs are introduced for water supply in Section 4.4, although these are also applicable to electricity or gas supplies. Probability analyses are described in connection with flood alleviation in Section 4.6.2, and the general concept of benefit transfer is mentioned in Section 4.6.3. The lower-bound concept features in several contexts, but is discussed in more detail in connection with a recreational-benefit estimate in Section 4.6.4. Transport is an example of a field where specialised software has much to offer. The use of alternative sources to value project benefits is described for power plants in Section 4.8. Many of these tricks of the trade can be applied more widely than their traditional ambit, and the reader is advised at least to skim through all the sections, even those describing fields that are not of immediate concern. Many projects involve several kinds of effect and therefore more than one section of this chapter will be relevant even to a single project.

This book cannot always replace good manuals and textbooks related to particular fields or types of project. The individual sections of this chapter discuss the main principles of each field, introduce concepts and terminology, give examples, and then refer to specialised works for further information and details.

Cost–benefit analysis – A practical guide
ISBN: 978-0-7277-4134-9

The question of whose point of view is represented in a CBA has been emphasised in Chapter 1 and elsewhere, as a distinction between different types of CBA. In some fields it may be necessary to conduct analyses from two or more points of view, because of the importance of *motivation*. This applies to projects whose users may choose whether to make full or any use of what the project offers. An irrigation project may be capable of delivering certain economic benefits to the nation, as revealed by a normal economic analysis, but it will not actually do so unless farmers choose to use irrigation in the way assumed. That will depend on whether or not the extra income or food justifies, in their eyes, the extra effort or risk; they will decide whether or how to use the facilities on the basis of their own private (usually informal or even unconscious) CBA. In studying such a project the analyst will usually do an economic CBA to find out whether the project is desirable or acceptable from the nation's economic point of view, and also a financial CBA for a typical farmer to find out whether it is practically viable. Similar checks may be necessary for many types of project, both to prove viability and to help estimate demand: Will drivers choose to use a toll road? Will patients use a proposed health service? Will parents send their children to a new sort of school? These subsidiary analyses are usually used for screening (separating viable from non-viable projects), while an economic analysis from the national viewpoint is used to determine absolute (yes/no) or relative (ranking) desirability. The subsidiary analysis will often be done in financial terms, if that is judged to represent (to a sufficient degree of precision) the factors deciding the motivation and action of the potential users. One significant variable in such an analysis is, of course, the fee or charge, if any, that users will have to pay for the service, and the same variable is also a major determinant of the financial balance of the agency or firm that will operate the project: overall project viability also requires that this body should be sufficiently motivated to play the part assumed, and should not collapse financially. (This is often discussed under the name of 'cost recovery'.) Determining or designing for project viability may thus involve a set of interlocking financial analyses. Demand analysis and forecasting is a complex subject, specific to each field, and is outside the scope of this book.

This chapter necessarily takes project types and purposes one by one, but a multipurpose project will of course involve valuation of several kinds of benefits, perhaps also of costs. For its economic optimisation and justification, such a project should be treated as one unit. Once it has been justified, however, an allocation of costs between the purposes may be necessary, for financing or for comparison with a competing project that offers an alternative for just one of the purposes. Allocation of the costs of a multipurpose project is a matter of arbitrary convention, and the best-known one is the so-called *separable costs – remaining benefits* method, which is explained in Box 4.1.[1] This method is usually used to allocate financial costs so that the agency promoting each separate purpose finds it worthwhile to participate in the multipurpose project. It can also be used to disaggregate such a project's economic CBA and give separate indicators for each purpose. In the example given in Box 4.1, the irrigation purpose produces less than a third of the benefits, but has relatively low separable costs, so it is allocated more than half the joint costs and still shows a single-purpose B/C far higher than the other purposes or the whole project.

Most CBA techniques are applicable equally to developed, industrialised economies and to less developed economies, but there are some matters that require more attention for developing countries. In the CBA context this term tends to mean countries with non-convertible currencies or restricted international trade, so that any analysis involving tradeable commodities (imports, import substitutes and exports) will need special care in the valuation of those elements, as described in Appendix B. The costs and benefits of projects in many fields may also need different treatment in less and more developed countries, because scarcities and priorities tend to be different, and because the database is often poor in less developed countries (e.g. they lack generalised tables of flood damage by building type, flood depth, duration, etc.). Many of the books and papers referred to in this book and listed in Appendix K are primarily related to less developed economies: this includes most books published by or for the World Bank, the Asian Development Bank, and United Nations and bilateral aid agencies,[2] as well as other books whose emphasis on developing countries is revealed in their titles or their author's affiliation.[3]

One topic not generally mentioned in the sections that follow is *demand forecasting*. This is a specialised task in each field, and outside the scope of this book, but its importance can scarcely be overemphasised. It is all too easy to design and then justify a project on the basis of an overoptimistic demand forecast, and then see it underutilised and, if re-examined retrospectively, judged to have been economically unjustified. Demand forecasts should be realistic, taking into account all relevant political and technical uncertainties, and should be linked into forecasts of the economy at large. Normally a CBA should use two or three alternative demand forecasts, such as a high-growth and

Box 4.1 *The separable costs – remaining benefits method*

Take, for example, a project comprising a dam with a hydropower station, which also supplies water to an irrigation scheme and gives flood control benefits. By the usual methods we estimate the total project cost, and the benefits associated with each particular purpose. We also design and cost hypothetical project variants that omit each purpose in turn. Each cost or benefit figure is a present value that includes all initial recurrent, replacement and decommissioning elements. We define:

- The *separable cost* for each purpose is the difference between the cost of the multipurpose project as proposed and what it would cost if that purpose were omitted; this includes not only the purpose-specific elements like the power station but also a share of the common elements like the dam and access road.
- The gross benefit for a purpose is the evaluated benefit of the relevant goods and services, or the cost of an equivalent alternative source, whichever is less.
- The remaining benefit for a particular purpose is that purpose's gross benefit minus its separable cost.
- The total joint cost for all purposes is the difference between the total project cost and the sum of the separable costs for all the purposes.

The principles are that:

- Each purpose should be allocated a cost that is not less than its separable cost and not more than its gross benefit.
- The joint cost should be allocated between purposes in proportion to their remaining benefits.

The following example shows how the principles are satisfied ($J =$ total joint cost, amounts shown in million dollars at 2003 economic prices).

Purpose		Hydropower	Irrigation	Flood control	Total
Gross benefits	A	133	75	30	238
Separable costs	B	103	38	25	166
Remaining benefits	$C = A - B$	30	37	5	72
Share ratio	$D = C/\sum C$	0.42	0.51	0.07	1.00
Joint costs allocated	$E = J \times D$	12	15	2	29
Total costs	$F = B + E$	115	53	27	195
NPV	$A - F$	18	22	3	43
B/C	$A + F$	1.16	3.4	1.11	1.22

a low-growth scenario, and these should form part of the sensitivity tests. An example is given in Chapter 5.

4.2   Agriculture

Agricultural projects are an example of those where several parties are involved between an investment (or the discrete injection of resources that the project represents) and the final output. Analysis of such projects is therefore described in some detail here because the approach can be used for other sorts of project that share this characteristic. Suppose, for example, that a project provides irrigation to an area which previously had only rain-fed cropping. The engineering element of the project is relatively straightforward, consisting of resources used to build canals, roads and drains and their structures, and then to maintain and operate them throughout the analysis period (this latter part must never be omitted). The costs of this engineering element can be estimated in financial terms, and converted into economic terms, in the usual ways.

The benefit side is not so simple. The immediate product is water provided to farmers' fields in certain quantities and at certain times that correspond (sometimes not very well) to the needs of the crops during the different seasons of the year. If it were possible to put an economic value on the water delivered to the fields, for instance per cubic metre, then the analysis boundary might be drawn there and the economic value of the water counted as the benefits of the project. In fact, however, one can only estimate the economic value of the water by considering what the farmers do with it. It is usual, therefore, to draw the analysis boundary so as to include the fields and the crops as well as the canals. Now the product to be valued is the harvested crop, and the costs include not only those mentioned above but also the labour, seed, fertiliser, herbicides, tractor-hours and whatever other resources the farmers use to produce the crop. There are now two distinct parties involved: the project operating authority, and the farmers. Sometimes two kinds of farmer, such as large estate farmers and smallholders, may have to be considered as separate parties. The economic analysis, done from the national point of view, will of course lump all these parties' costs and benefits together. But, unless the operation is totally controlled by the project authority, with the farmers as its employees, the analysis needs also to ensure that the farmers will actually grow the crops assumed, which they will only do if they perceive a favourable private CBA for each crop. It is usually also desirable to check that the project authority will break even financially, or to estimate what subsidy from the state and/or fees from the farmers it will need to receive. We therefore have not only the overall economic analysis of the project (defined to include the fields and crops) but also the private or financial analysis for one or more typical farmers and for the project authority.

For this reason we do not lump the costs together, but treat the farmers' costs quite separately from those of the entity that builds and operates the infrastructure. The key to this is the subtraction of farmers' costs from the gross return to estimate the net return. In financial terms (which are those affecting the farmer for all elements except family labour and owned land) the *gross return* is the money he gets from a trader or other buyer when he sells his crop, and the *net return* is the gross return minus production costs. (In this section these terms will be used, but the reader should be prepared to find others in other books and reports; for example, *gross margin*, which is sometimes equivalent to what is here called net return, but at other times is an intermediate subtotal after the subtraction of some of the costs and before the subtraction of others.) When the farmer borrows money, or pays rent, or shares part of his crop with a landowner, the farm-level analysis becomes quite complicated. This section presents a simplified example, based on an actual feasibility study, which can serve as a framework into which refinements can be inserted for specific projects.

In order to assess the attractiveness of individual crops to farmers, as well as a step towards building up the project CBA, we start by computing the net return for each crop, per unit area of land. It is often useful to note the return to a unit of *family labour* also, because the farmer's choice of crop may be influenced by return to land, or return to labour, or both, depending what he perceives as scarce. This computation is called a *crop budget*. Table 4.1 shows one of many crop budgets from the analysis of a particular irrigation project. This particular crop budget is for maize and is done for the with-project situation at constant 1998 financial prices. In order to use a consistent format, the table provides lines for several items that are not applicable to this crop but are needed for others in the full analysis (e.g. gross returns for by-products and sacks). The table indicates that, with no financial payment by the farmer for land, irrigation water or family labour, the farmer would see a net return of 1392 pesos for every hectare planted to maize, and every person-day of family labour would bring in 23 pesos. This is considered sufficiently attractive, relative to other

Table 4.1 A simple financial crop budget

Crop budget for 1 ha of maize

Farming system	Smallholder irrigation
Situation	Future with project
Prices	Financial
Currency	Pesos at 1998 constant prices

Item	Unit	Quantity	Price	Value
Gross return				
Main product	t	4.0	565	2260
By-product	t			0
Losses	t			0
			Gross return	2260
Production costs				
Seed	kg	20	0.20	4
Fertiliser:				
N	kg	110	2.86	315
P_2O_5	kg	60	2.30	138
K_2O	kg	90	2.39	215
Agrochemicals	peso	40		40
Machinery:				
Tractor	h	2.0	25.00	50
Combine	h	0		0
Animal power	peso	10		10
Sacks, etc.	peso	0		0
Hired labour	person-day	8	12.00	96
Family labour	person-day	60	0.00	0
			Total costs	868
Net financial return per hectare				1392
Net return to family labour (pesos/person-day)				23

uses of land and family labour, to motivate the farmers to grow maize under irrigation. The data in the table refer to one crop season, and in this case there was only one maize crop per year.

Similar financial crop budgets were prepared for other crops, and it was considered likely that smallholder farmers, having the use of 5 ha of land per farm family, would use it to grow 3 ha of maize, 1 ha of cowpeas, and 0.5 ha each of cabbage and tomatoes, using the proposed supply of irrigation water. The list giving the proportions of the various crops is called a cropping pattern. Now the analyst computes *a farm budget* for this particular farming system, as shown in Table 4.2. This refers to one year, and in this case there is only one cropping season per year for each of these crops. The net return to land for maize, 1392 pesos/ha, appears in the third column above the corresponding

Table 4.2 A typical farm budget

Farm budget for smallholder irrigation

Situation	Future with project
Prices	Financial (projected)
Currency	Pesos at 1998 constant prices

Crop	Area for this crop: ha	Net return: pesos		Family labour	
		Per hectare	Per farm	Per hectare	Per farm
Maize	3.0	1392	4177	60	180
Cabbage	0.5	2270	1135	110	55
Tomatoes	0.5	2630	1315	125	63
Cowpeas	1.0	519	519	31	31
Total	5.0		7146		329

Farm net return per hectare	1429
Farm net return per person-day of family labour	22

figures for the other crops, and the fourth column shows the return to the 3 ha of maize that the farm grows (i.e. 4177 pesos). Adding the corresponding figures for the other crops shows that the net return for the whole farm is 7146 pesos, which is 1429 pesos/ha. The right-hand columns in Table 4.2 give the overall financial return to family labour, and this shows that the motivation for irrigated farming as a whole is sufficient. (It is not evident from this table why the farmers would bother to grow cowpeas, which give a lower return per hectare and per person-day than the other crops. The explanation was that labour was limited in total availability and that cowpea was a food that people wanted. Any analysis must be realistic and must recognise that farmers choose their crops for reasons other than financial returns, such as taste, habit or social custom. The analyst draws up the cropping pattern in the light of farmer interviews and general experience, as well as of the computed returns.)

Incidentally, this financial farm budget also serves to guide decisions about what water charges could be levied by the irrigation scheme authorities without removing farmers' motivation to engage in irrigated cropping. Other farming systems, with different farm sizes and cropping patterns, might be included in the overall mix of the project, in which case each would be analysed as in this example.

It is important to note that in this financial analysis the hired labour was costed at the financial price that the farmer would actually pay to labourers, which was determined by the local labour market and by legislation, while family labour was costed at zero because the farmer does not pay money for it. The latter is mentioned in the tables only so as to compute the financial return to it, and to set up the framework for the economic analysis that follows.

The next stage is to conduct the economic analysis, from the point of view of the nation. The same framework of crop budgets is used, but the financial prices are replaced by economic prices. For inputs such as fertilisers and tractor-hours this involves normal shadow pricing, taking account of any elements of imported factors or of taxes. For hired labour the economic price was much lower than the financial one, as that was kept up by legislation and the opportunity cost of rural labour was low. The family labour, which was valued at zero in the financial analysis, now has a finite economic value. Table 4.3 shows the economic crop budget for maize, which can be compared

Table 4.3 An economic crop budget

Crop budget for 1 ha of maize

Farming system	Smallholder irrigation
Situation	Future with project
Prices	Economic
Currency	Pesos at 1998 constant prices

Item	Unit	Quantity	Price	Value
Gross return				
Main product	t	4.0	544	2176
By-product	t			0
Losses	t			0
			Gross return	2176
Production costs				
Seed	kg	20	0.20	4
Fertiliser:				
N	kg	110	3.07	338
P_2O_5	kg	60	2.43	146
K_2O	kg	90	2.54	229
Agrochemicals	peso	40		40
Machinery:				
Tractor	h	2.0	28.00	56
Combine	h	0		0
Animal power	peso	10		10
Sacks, etc.	peso	0		0
Hired labour	person-day	8	5.00	40
Family labour	person-day	60	5.00	300
			Total costs	1162
Net economic return per hectare				1014

with the financial one given in Table 4.1. Table 4.3 shows that the economic net return on 1 ha of maize is 1014 pesos (at 1998 economic prices).

Now the economic net returns for all the crops and all the farms are calculated and combined, to give the annual economic net return for the whole project. This can be done via economic farm budgets, and then multiplying by the number of farms of each type in the project; or it can be done directly from the per-hectare economic crop budgets to project level, merely multiplying each crop's net return per hectare by the number of hectares of that crop grown in the project. The worked example in Appendix I, which carries this example further, uses the former method in order to facilitate a certain sensitivity test.

The costs of an agricultural project will often contain land costs as an important element. The valuation of these is discussed in Section C.3 of Appendix C, and is often dealt with by judicious definition of the without-project situation.

The annual economic net return calculated in this way for the with-project situation is now set alongside the corresponding figure for the without-project situation, which in the case of a new irrigation project is rain-fed farming, but in the case of a rehabilitation project represents continuing substandard irrigation. The difference gives the incremental economic net return at project level, which is the measure of the project's economic benefits. In most projects it has to be recognised that the incremental returns will not flow at the full annual figure from the first year of operation, but will build up gradually to that figure over a few years, as in the example in Chapter 3 (Table 3.2, page 47, showed annual benefits of 10, 30, 90 and 160 units in the first four years, and then a steady 200 units/year). In an agricultural project this might be due both to staged completion of the infrastructure and to the time taken by farmers to learn the techniques of irrigated agriculture.

The stream of incremental economic benefits for all years of the analysis period is now set against the incremental project-level costs, for the canals, roads, operation and maintenance, and so on, all calculated in economic terms. The project's economic cost–benefit indicators are calculated as explained in Chapter 3.

This agricultural example illustrates several points of interest, which can often apply to other sorts of project also. They are:

- In the economic analysis the agricultural production costs (labour, fertilisers, etc.) were not counted with the construction and operation costs but were subtracted from the gross returns to give the net returns, in the economic crop budgets. In effect they were counted as *negative benefits* rather than as costs. It is evident from the way the CBA indicators are calculated that this will make no difference to the NPV or IRR, but will give a different B/C from that which would be found by counting them as costs. In fact it is conventional to analyse agricultural projects as described here, largely because the crop budgets are conceptually a way of valuing the water delivered to the farms, and it is that delivery of water that is perceived as the benefit to set against the engineering costs.

- When this sort of multilevel treatment of benefits is used, care has to be taken with terminology because several subtractions take place. In the terminology used here, we first subtracted agricultural production costs from the *gross returns* to get the *net returns*, and then subtracted without-project net returns from with-project net returns to obtain *incremental net returns*, which are a measure of the benefits, year by year, for the project. Finally, in the project-level economic analysis, we subtracted annual costs from annual benefits to get annual *net benefits*, as that term was used in Chapter 3. The words *net* and *incremental* both imply subtraction, but at different stages. Some people use these terms, and the word *margin*, differently from the way they are used in this section, which is not a problem provided that the words are clearly defined and consistently used. (Another set of terms for crop budgets is: *gross output* (crop yield × price) minus on-farm consumption gives *gross income*; this minus cash costs gives *cash margin*; this minus family labour cost gives *net margin*.)

- A set of financial analyses for distinct parties, in this case the typical farmer and the scheme authority, is needed as a complement to an economic CBA. The financial analyses for these crucial parties show whether the project will actually work as planned, i.e. whether it is feasible (viable, sustainable), in the sense that the parties would actually be sufficiently motivated to play the roles assumed for them, and would not lose interest or go bankrupt. This financial and motivational viability is just as necessary as technical feasibility (canals that flow, bridges that do not collapse, etc.). Meanwhile the economic analysis shows whether the project is economically desirable (justifiable, acceptable), i.e. whether it is worth doing. Both analyses are needed, as it would not be clever to proceed with a project with an attractive economic indicator if it was not going to function because one of the key parties had no good reason to participate.

- These financial analyses for the individual parties give guidance as to what financial payments will be needed to make the project work; for instance, water charges paid by farmers to scheme operators, or subsidies paid by the government to the same operators. The financial farm budget shown above omits such charges. After making that analysis and another, also without water charges, for the scheme operator, one might decide on a water charge amounting to 300 pesos per 5-ha farm per year. One could then re-work both financial analyses to include these payments, which are called *cost recovery*, so as to show that the farmers can afford to pay and that the operator does or does not now need a government subsidy as well. But all this would not affect the economic analysis, which includes both farmers and operators. In that analysis the water passing from canal to field is a physical transfer within the defined analysis boundary, and the water charge money flowing the other way is a transfer payment, not to be counted in an economic analysis that values its benefits as net agricultural returns on crops rather than as a value placed on water.

- A further lesson from the above simplified example, a matter of practical technique rather than one of principle, is that a systematic approach to the financial analysis will facilitate the subsequent economic one. For instance, the financial crop budget was set out in a spreadsheet table that contained a variable for the price of family labour, although the price was necessarily zero in that analysis, so that the same table format could afterwards serve for the economic crop budget. Analysis of a real project will often involve ten or more crops, each to be analysed at financial and at economic prices for the with-project and the without-project situations, so a tidy layout that can be replicated dozens of times for all these cases will ease the analyst's burden and help to eliminate mistakes in the calculation. It will also help to make the analysis transparent to readers and reviewers of the report. The same reasoning underlies the systematic statements about situation, prices and currency at the top of each of the above tables.

Many of the features of agricultural project CBA as described here are applicable to other sorts of project where the physical product (e.g. water, electric energy, better transport) is difficult to value in its physical form. The trick is to define the analysis boundary in such a way that some other commodity (tons of maize in the example) is valued in economic terms, and to work backwards by way of the technical and financial and economic linkages (crop yields, water requirements and production costs) to the original product (water in canals). As well as solving the problem of economic valuation, this enables the analyst also to verify the motivational feasibility of the project. The analogies are fairly obvious: for water take the provision of a bridge, for farmers take drivers, and for water charges take tolls: then we have the analysis of a toll bridge.

For more detailed guidance the reader should consult Gittinger (1982), a thorough and practical guide covering many aspects, including livestock, and presenting many examples; it also has an extensive glossary. The more recent handbooks and guidelines produced from time to time by the financing agencies (such as the World Bank and the Asian Development Bank, as well as various countries' bilateral aid departments) are also useful sources.[4] Estimating the yield differentials and other benefits in different situations can be assisted by using published quantitative linkages found in other places (e.g. Finney, 1996), especially when the project includes land improvements such as field drainage or land levelling. Such relationships are effectively *dose–response functions* like those mentioned in Appendix C, and their use in places remote from the project site is an example of *benefit transfer*.

Forestry projects tend to have particularly long time horizons and long delays between investment and the flow of benefits, because trees take so much longer to reach harvest than most other crops. The benefits also include environmental ones, such as improvements in the carbon dioxide balance, which are difficult to value. For both these reasons, forestry is a specialist field within CBA. Arguments are often heard for using lower discount rates for forestry than for other sorts of project, and the UK Government has in the past done so, but it is fundamentally irrational to have different time preference rates for different things. A better approach is to use a general discount rate but take care to include all benefits, and to apply sustainability criteria to competing projects as well as to forestry ones.[5]

4.3 Health projects and programmes

Applying CBA to health matters is not easy, but in this sector like any other there are scarce resources to be allocated and therefore decisions to be guided. In making decisions about the provision of healthcare measures, including preventive ones, there are three main classes of costs:[6]

C1: Direct costs to the health service, i.e. those of hospitals, equipment, drugs and medical staff.

C2: Quantifiable costs to patients and their families, such as their lost production, or the extra cost of bringing food to a hospital instead of eating at home.[7]

C3: The intangible cost to the patients and their families due to their undergoing the treatment or receiving the care.

The benefits can be evaluated in various ways, and grouped as follows.

Health aspects:

E1: Reduced *morbidity* (incidence of disease), measured in practical units related to a particular disease or condition (e.g. disability-days saved).

E2: Reduced *mortality* (avoidance of death or of early death).

Economic benefits, in three classes mirroring the three kinds of costs:

B1: Direct benefits to the health service, i.e. reduced future health-care costs because the project makes people healthier.

B2: Benefits to the patient, i.e. production gained because people are fit and able to work longer, or sooner, or in greater numbers.

B3: Intangible benefits, i.e. the monetary or economic value of the reduction in pain, grief and suffering due to the project.

Value of health improvement itself, independent of economic consequences:

S: arbitrary numerical scales (1 to 10, 1 to 100, and similar)

W: willingness-to-pay and willingness-to-accept-compensation measures (see Appendix C)

U: utility measured by *quality-adjusted life years* (QALYs) or some similar measure.

Many of these parameters are very difficult to quantify, or need a great deal of data and analysis effort, and so techniques have been developed to produce useful, if crude, decision guides that use only some of them. The first category of methods is a particular kind of *cost-effectiveness analysis*. The cost measure is often just the costs that can be fairly easily quantified, C1 and C2, ignoring C3. Alternatively, a kind of net cost to society can be estimated, corresponding to $C1 - B1$ or, more subtly, $C1 + C2 - B1 - B2$. The ratio of this cost to the effectiveness is used to decide between alternative courses of action, effectiveness being quantified in terms related to the immediate objectives of the project, corresponding to E1 or E2 above. For example, the measure might be the number of cases prevented or life-years saved, or numbers of children immunised (although this measure is actually only an intermediate parameter on the way to the end result).[8] This technique is limited to comparisons between projects or programmes with closely similar objectives, as only one consequence measure can be used; the method cannot, for instance, deal with both mortality and morbidity changes at the same time.

A wider but more data-intensive set of methods is sometimes called *cost–utility analysis*. Here cost is quantified more or less crudely as before, but the effect of the project is quantified by a measure such as quality-adjusted life years (category U in the above list). Box 4.2 describes some of the indicators used. Methods like this can to some extent be used to make comparisons between dissimilar projects, in that the effectiveness measure can bring different sorts of health improvement to a common denominator.[9]

Box 4.2 Some measures of health project effects

There are various measures of the effectiveness of a medical or healthcare intervention. The simplest is *years of potential life gained (YLGs)*. For one hypothetical patient this is the difference between expected life duration with and without the intervention; for the whole project it is the sum for all patients affected, and if patients are treated in different years the figures for the years can be discounted to obtain a present value of YLCs. This measure primarily takes account of mortality changes, but not morbidity improvements.

Refinements generally consist of weighting the years of life gained in various ways. *Healthy years of life gained (HYLG)* takes direct account of mortality and morbidity, in that the years of life remaining to the patient after the intervention are weighted according to his relative health or disability. Another measure also applies subjective weighting by age group (*disability-adjusted life years*, or DALYs). The most subtle of such measures is *quality-adjusted life years (QALYs)*, where the years of extra life are weighted according to the quality of life on some scale; a healthy year counts as 1 and a year affected by suffering or infirmity as some fraction of a year. In many countries DALYs give the most useful measure.

A more sophisticated, but even more difficult and data-intensive, approach is to apply not cost-effectiveness or cost–utility analysis but CBA to guide the decision. In principle this can give indicators to guide decisions between very different health projects, the outputs of which cannot be brought into a common effect measure, or between health projects and those of other sectors. The simplest sort of health-sector CBA uses only the tangible economic cost and benefit categories C1, C2, B1 and B2 in the above list (C3 and B3 are usually omitted as being too difficult to define and measure). This is not entirely satisfactory, however, because B2 is a very limited and incomplete measure of the benefits of some sorts of health project. In its basic form it omits much of the benefit due to improved health, such as that affecting people whose main activities do not take place in the money economy and thus do not get measured (e.g. people engaged in housekeeping or subsistence farming). Some of these omissions can be remedied by adding in specific estimates, for instance using the opportunity-cost approach. Sometimes a further category is added, namely an estimate of the full non-production benefit (W in the above list). This has to be valued by the willingness-to-pay and/or willingness-to-accept-compensation methods, or others mentioned in Appendix C. For instance, if there is a market in healthcare its prices can give some indication of willingness-to-pay, while actual compensation payments, whether agreed with victims of particular health conditions or imposed by court settlements, can give a guide to willingness-to-accept-compensation (usually higher than willingness-to-pay). Adding in W alongside B1 and B2 does not constitute double-counting, provided that the estimate of W is designed to exclude production losses. W represents an approximate way to put a value on B3. In particular cases, the value of a statistical life may be included in the analysis.

Valuation of healthcare and public health benefits should be linked to the study of demand, which may be sensitive to price as well as availability to particular social groups and to style of delivery. As mentioned at the end of Section 4.1, it may be necessary, in order to prove a project's viability, to analyse its impact on particular groups of people who might choose to use its services or not, according to whether the benefits (of all kinds) that they perceive and experience outweigh any costs to them.

Many of the techniques and concepts mentioned in Appendix C may be used for CBA in the health sector, including the human capital approach, and the technical links between the project activities and the various approximately valued outcomes have to be quantified by use of dose–response functions, often from epidemiological models but also including sociological determinants of the effectiveness of interventions. Factors and relative values may be borrowed from other places (benefit transfer), but only with great caution.

It is evident that even a relatively crude analysis of a health project, with its inevitably complex set of linkages between inputs and final objectives, will be complicated and potentially controversial, and will need a great deal of data and analytical effort and skill. To obtain usable results, with realistic analysis costs, requires drastic simplifications, and often it is advisable to structure the decision so that only cost-effectiveness analysis is needed, not CBA. The more sophisticated techniques, especially QALYs, are normally only used in rich countries, and adapting results from earlier studies (benefit transfer) is useful between similar places in such countries. When a full CBA is done, the indicators often show very favourable results, especially in places where general health levels are poor and simple medical interventions can achieve dramatic results. The World Bank handbook gives an example of a CBA for an immunisation programme in a developing country (with admittedly crude assumptions, but ones that probably underestimate total benefits), resulting in an EIRR of 98% and a B/C ratio of 4.8.[10] Experts in the field might say it was obvious anyway that such a programme was very worthwhile, but there remains some value in putting a rough figure on its merit, when it competes for resources against other urgent imperatives such as education or defence. The same example shows how, alongside the rough absolute CBA, one can do a cost-effectiveness analysis to select the type of immunisation campaign, and that requires less shaky assumptions.

This book cannot do more than describe these few generalities of this complex subject. Readers requiring more guidance on CBA or related techniques in the health sector should consult the *Journal of Health Economics*, *Health Policy and Planning* or the *International Journal of Health Planning and Management*, or one of the references listed in the Notes.[11]

4.4 Water supply, sanitation, public health

Projects in these interrelated fields have both health and engineering aspects. The costs are relatively straightforward, and can be estimated and, if appropriate, shadow priced in the usual way. The benefits are more difficult to analyse. Water supply is metered and paid for by volume in some places, in which case there are tariffs that quantify the financial benefits of increased water supply from the water company or agency's point of view. Sewerage or other waste-disposal services are seldom metered or paid for by quantity. In any case, the price paid for such services is seldom a full measure of their economic value to the nation or society as a whole, because the provision of the service to individual consumers or households yields benefits to others in the form of improved general public health (especially reduced disease transmission) and better environmental or aesthetic conditions.

The traditional response to these difficulties is to restrict oneself to a cost-effectiveness analysis. Typically, for water supply, the *average incremental cost* of water (sometimes called the *unit reference value*) is calculated and compared with that of existing or alternative projects or water sources. Most projects have not only an uneven stream of costs over their lifetimes but also a non-uniform pattern of water volume supplied, beginning with only part of the capacity used and building up as demand grows. The average incremental cost is computed by discounting the successive annual costs (in financial or economic terms) to give the present value of costs, and also discounting the successive annual volumes of water supplied, at the same rate, to give the discounted total volume of water. (This is probably the most common example of discounting a quantity which is neither money nor even measured indirectly in monetary units.) The simplest use of this procedure is to compare the financial average incremental cost with the proposed tariff, giving an indication, incidentally, of the financial viability of the project without subsidy. It can be convincingly argued that the full economic benefit per unit water volume must be greater than the financial tariff, so if the average incremental cost is less than the tariff the project must be economically acceptable. This is a crude lower-bound calculation that will justify a few projects, but cannot be used to reject a project or to rank different projects.

More refined analysis is possible.[12] For water supply it depends mainly on estimating the willingness-to-pay for piped water. This varies considerably between consumers and with the quantity of water used. The bare basic needs for water are very low, as little as 5 litres per person per day for drinking and cooking. But if water is available at an affordable cost (cost may be financial or in terms of effort and time to fetch water from some distance away), people often use 50, 100 or even 200 litres per person per day. The demand varies with price, the demand curve typically sloping at a 0.2% to 0.8% reduction in quantity demanded per 1% increase in price, over most of the range.[13] Even where supplies are metered, people usually pay as a tariff much less than their willingness-to-pay (the difference being their consumer surplus). By investigating what people without piped supplies are paying for water (to vendors, well owners, or in terms of family labour fetching water), one can make a rough estimate of the demand curve and hence the willingness-to-pay, from which the full economic value of the benefits of water supply can be estimated. They may be several times the tariff, so such an analysis can provide a sound economic justification for a water-supply project that, under the more traditional approach, would have been implemented on the grounds of basic needs or just on the assumption that water supply must be provided. In this way the relative merits of different projects can be rationally assessed.

Where the without-project situation involves poor quality water or water fetched from a distant source, willingness-to-pay and hence project benefits may reflect not only the avoided cost of fetching water (often unpaid time of women and children) but also the avoided cost of boiling it if the project water is potable.[14]

A significant by-product of a willingness-to-pay analysis is a yardstick against which tariffs can be reviewed. This in turns links to demand forecasting and demand management. A realistic tariff will provide the motivation for consumers to use water appropriately, which in many circumstances is a better and safer way of limiting demand than rationing by the more usual methods of load-shedding or part-time or low-pressure supply.

Both water supply and waste-water projects or policy changes usually have implications for public health; indeed, that is often the proponent's main aim. Willingness-to-pay will generally fail to capture the collective benefits of public health improvements and improved environmental conditions, because these are to a large extent public goods (see Box C1 in Appendix C, page 174). Some of them may be valued as health improvements under categories B1 or B2 as given in Section 4.3.

Readers who wish to know more will find Klümper (1995) or Merrett (1997) useful. Willingness-to-pay for water and sewerage services is discussed in Green *et al.* (1993), and an appraisal of sewerage schemes in general is given in Green *et al.* (1989). Water supply schemes are well covered in an Asian Development Bank paper by Whittington and Swarna (1994). An example of CBA concerning water pollution is given in James (1994).

4.5   Education
CBA can be used to guide decisions on resource allocation in education; for instance allocations between primary, secondary and tertiary

education, between different sorts of schools, or between subjects. The starting point is usually financial, although allowance can be made for effects (usually called 'externalities') that are not directly valued in monetary terms. On the cost side there are the direct costs of providing education, plus the opportunity cost of the student's time, usually measured by lost earnings (so that a teenage student's time is valued at a lower price than that of someone who returns to education in mid-career). The benefits are usually valued by observing the differential lifetime earnings between people with differing amounts of education behind them. The underlying theory often goes by the name of *human capital*. Large amounts of earnings data are assembled, covering people's whole lifespan, and then smoothing and regression analyses are done to separate out the effects of education for different periods or sorts of education.[15] Once the effects of experience on earnings have been statistically separated, along with other influences if the database permits, the average effect on lifetime earnings of any particular type or amount of education can be deduced. This remains only a partial estimate of the benefits, because earnings do not capture the whole differential productivity and usefulness of educated people. Some goes to their employers and to society in general: the measure shows only private benefits, not social benefits. The earnings measure also fails to put any value on the enjoyment and satisfaction experienced by the student or educated person, which is significant (and in many cases a major motivation for seeking education). Another problem with the earnings measure is that it suffers from serious gender bias in that women have a different propensity to engage in paid work from that of men.[16] The earnings measure may, however, overstate educational benefits, because some earnings differentials may be attributed to education that are really due to something else, such as differences in innate ability, family connections, social class or ethnic factors. The whole method is very approximate, and the results must be interpreted with caution.

Estimating the public or social cost–benefit relationships (those from the government's and from the whole community's points of view, respectively) is even more difficult than estimating the private returns. On the cost side the costs that are met by the government or community instead of by the student (or her family) can be easily identified and added into the analysis, but on the benefit side things are more complex. Education will often cause increases in tax revenues to a government which are very significant compared to the government's outlay. It may also have extensive social benefits. For instance, an intervention specifically affecting the education of girls may result in changes in family size and population growth rate as well as significant health benefits.[17] Education may also have social disbenefits when raised expectations are not fulfilled (e.g. disgruntled street-sweepers with doctorates).

Despite all these difficulties, specialists in the field do make estimates of CBA indicators for education. They usually quote IRRs, which tend to range from 5% to 50%. Private rates reported for many countries are higher than social ones, although this is sometimes because only so-called partial social returns are calculated, adding government costs to private costs but not adding any non-private benefits. The less developed countries tend to show higher rates of return to education than more developed ones, because educated people have more advantage in the labour market when they are scarce. There is evidence that IRRs are sometimes used inappropriately when NPVs would be better guides to decision-making. For government decisions it is important to trace all tax consequences of education.[18] Returns vary widely between different sorts of education, and may be particularly high for the training of artisans.[19]

Because of the difficulty of estimating benefits in the education field, it is, of course, worth trying to formulate decisions in the least-cost or cost-effectiveness form. This can help with allocating resources between different ways of providing some specified type of education, but not, of course, with allocations between different types.

For further guidance and information the reader is referred to specialist publications included in Appendix K, such as Psacharopoulos and Woodall (1985), Psacharopoulos (1987; especially the papers therein by McMahon and by Woodall), Johnes (1993), Mills and Lee (1993), Cohn and Johnes (1994), and Brent (2009). Belli *et al.* (2001, Ch. 8) give practical guidance and examples, but mostly limited to financial benefits.

4.6 Flood and coastal protection, environment, recreation

4.6.1 *General*

These topics are linked because many projects that aim at reducing the risk (probability or likelihood) of flood or storm damage also have environmental impacts, and anyway such risks are part of the environmental conditions in any place. One of the main ways that coasts and many rivers are used is for recreation, so recreation values are relevant for many projects in these categories.

Throughout this section the term *damage-loss* is used to indicate the sum of costs resulting from a flood or storm event Some of this is straightforward damage, such as a building's walls being weakened and disfigured by flooding, or furniture and equipment being damaged so as to need replacement or extensive repair, and some is consequential loss or damage, such as the lost production from a factory that is out of action for several weeks after flooding.

There is sometimes a clearly quantifiable financial value of environmental improvements, for instance when a recreational facility is involved and users will have to pay to enjoy an enhanced environment. A financial analysis must of course include such payments, but need look no further. In an economic analysis, however, there are usually no financial flows corresponding directly to most of the environmental effects, yet these are real gains and losses from the nation's point of view and must be taken account of in any decisions. There are two main ways to tackle this:

- The environmental impacts, whether costs or benefits, are given economic values; this is difficult and imprecise but it can be done.
- Alternatively, the decision and its guiding CBA can be formulated in such a way that valuation of the environmental aspects is not necessary.

The second option may seem like an evasion, but often it will promote good decisions by concentrating the analysis on the aspects it can best illuminate. An example is given in Section 4.6.4, but the intervening sections deal with the more difficult case where environmental improvements, as well as any changes in risk of damage-loss, are explicitly valued in an economic analysis. When all or most of the loss avoidance and environmental effects of a project have to be valued, the methods described in Appendix C are used. The sections that follow illustrate how those general methods and principles can be applied to particular sorts of project. They describe flood alleviation, coastal defence and recreational benefits, respectively, but should be read together because many elements are common or similar. Evaluation of environmental aspects is discussed in Winpenny (1991) and Harberger and Jenkins (2002, Part II), as well as the specialised references in the notes to this section. A particularly thorough account is given in Hanley and Barbier (2009).

4.6.2   Flood alleviation and coastal protection
The valuation of benefits for flood alleviation projects is complicated by the amount of data needed, but the basic concepts are quite simple. The

project reduces the frequency and extent of damage-loss due to flooding, so we need to put an annual value on the changed probability situation. As explained in Appendix A, the annual *cost-of-risk*, or annual equivalent cost of damage-loss, for an event of known annual occurrence probability and known damage-loss is:

cost-of-risk = probability of event × damage-loss caused by event

In any situation, such as the with-project and without-project situations that define any particular project or variant thereof, there is usually a whole range of possible events with very different damage-loss consequences, from a mild flood that is quite likely to occur in any one year but causes minimal damage, to a catastrophic flood that has a minuscule probability of occurrence but would cause massive damage if it did occur. As illustrated by the example in Appendix A, the above multiplication of probability by damage-loss must be done for the whole of the probability range or spectrum, either by dividing it into slices and formulating a typical event for each slice, or by estimating the damage-loss/probability function or graph and finding the area under it.

Thus the main effort of a CBA in these fields is the estimation of the damage-loss/probability function, where damage-loss is the general term that includes the economic valuation of all consequences of an event, including indirect losses and costs as well as the more obvious damage to people, buildings and infrastructure. This is usually done in a number of steps, according to the nature of the project and the damage. In the bridge example in Appendix A there were just two steps: damage-loss was related to flood discharge by engineering considerations, and flood discharge was related to probability by normal hydrological analysis. For river floods affecting buildings and land, the sequence might be:

(a) estimate the probability (frequency) of different high river discharges (the flood frequency curve or probability/discharge function)
(b) estimate the water levels relating to each discharge
(c) estimate the damage-loss consequent to each water level (methods of doing this are discussed below)
(d) put the above three functions together to produce the damage-loss/ probability function.

If damage-loss is related primarily to area, as with damage to agricultural crops or land, there may be an intermediate step between (b) and (c); after the discharge/level function (hydraulics) a level/area function is estimated (topography), and then an area/damage function (agricultural economics).

Once the annual equivalent damage-loss functions have been estimated for the with-project situation and the without-project situation, they can be discounted over the analysis period and the one subtracted from the other to give the incremental benefit attributable to the project. It does not matter whether the subtracting or the discounting is done first, and indeed it is often illuminating to do either or both for individual slices of the probability range, before they are added together, so as to show what sorts of flood are most important for the generation of benefits. Even if the integration over the probability range, and the with-project minus without-project subtraction, are done digitally, it is worthwhile to show the damage-loss/probability function graphically, as in Figure 4.1. Shading the areas under the function for the with-project situation (the residual damage-loss) and the area between the curves of the two situations (the averted damage-loss) conveys a great deal of information in a compressed and digestible form. In particular, it shows what part of the probability range (what sorts of flood) contribute most of the benefits. For urban flood

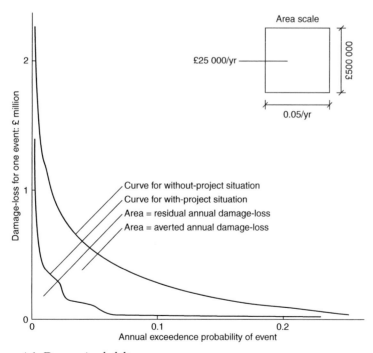

Figure 4.1 Damage/probability curves

alleviation measures, most of the benefits tend to arise from reducing the severity of flooding from middle-probability events, such as return periods of 2 to 50 years (annual exceedence probabilities of 0.5 to 0.02), rather than from more dramatic extreme floods of a return period 100 years or more (0.01 and less). The number of flood sizes (events) analysed should be at least three, perhaps up to eight, for reasonable precision. The damage-loss/probability function is necessarily asymptotic to the damage-loss axis (a very large and rare flood would produce enormous damage), and it is often almost asymptotic to the probability axis too (small floods such as would be exceeded every year would produce negligible damage-loss); these two conditions are a considerable help in drawing a realistic damage-loss/probability curve with limited data. Without such guidance, engineers often tend to spend scarce resources on excessive levels of protection.

Usually the whole cost-of-risk estimate is repeated for several alternative project designs that offer different levels of flood protection, so that one can then choose the economic optimum (by the maximum-NPV rule, of course, as the alternatives are technically mutually exclusive ones).

A major part of the workload of someone conducting a CBA for a flood alleviation scheme is the physical quantification and then economic valuation of damage-loss. In some countries, extensive research leads to standardised unit loss figures for particular types of building or property. An example is the highly developed methodology in Britain, mostly the work of the Flood Hazard Research Centre (FHRC) at Middlesex University. This has been published in a series of manuals, as well as numerous papers and conference contributions. The *Blue Manual* [20] was published in 1977, giving a detailed procedure for valuing damage-loss resulting from flooding of residential, commercial, industrial and agricultural areas. In 1987 this was updated by the *Red Manual*, [21] which also refined and extended the treatment of indirect losses. The detailed tables of damage-loss figures for the various types of property and degrees of flooding have, since then, been periodically updated by the FRHC. In 1992, the FHRC issued the *Yellow Manual*, [22] which was primarily concerned with coastal defences but also refined the methodology of the first two manuals. The team of researchers at the FHRC has continually developed and refined their categories and methods as well as the unit values, and have arrived at a well-tried and sophisticated methodology for economic valuation of most sorts of damage-loss associated with floods, storms and, by extension, lightning and such events. Their work up to 2003 was used in the preparation of the *Multi-Coloured Manual* and associated handbook, since updated.[22] The hydrological part of the task, estimating a flood probability function in the UK, can be guided by the *Flood Estimation Handbook*, which is periodically updated and available on the internet.

The main subdivisions of damage-loss due to a flood are direct and indirect, the direct being more in the nature of damage and the indirect one more in the nature of loss, although there is no sharp distinction between damage and loss.

Direct damage includes the cost of replacing or repairing buildings, furnishings, equipment and other physical objects that are damaged by the water. For a particular type of building the damage depends heavily on the depth of flooding and its duration. In the UK, for example, a particular FHRC table gives monetary damage estimates for a modern bungalow occupied by people of social class C2, after flooding to a depth of 0.2 m above the ground-floor level for less than 12 hours: it covers building fabric damage and inventory damage separately. The inventory damage figures have been painstakingly built up from individual items such as a sewing machine and a dining chair; the classification by occupant's social class (using the advertising industry's classes) serves to determine what sort and quantity of vulnerable items a dwelling is likely to contain, and at what levels above the floor they are located. The regularly updated tables embody a vast amount of information that an analyst would never have time to collect directly for a particular flood alleviation scheme; it is only necessary to quantify what numbers of dwellings of each type are flooded, and to what depth and duration, for each probability level and flood protection scenario, and to multiply by the unit values from the FHRC tables. The tables cover a wide range of building types and land uses, for instance the direct damage-loss per square metre of a shoe shop, or a pub without a cellar, all in terms of depth and duration. Computer programmes are available to carry out the very extensive computations involved. For agricultural areas, the FHRC system covers crop losses as well as damage to buildings and equipment.

The estimation of indirect damage-loss is another complex procedure. This category includes the value of production lost during some period after a flood has subsided because a factory's equipment has been damaged, and this has to be added to the direct cost of replacing the equipment. Care is taken to count only losses from the nation's point of view: if production is lost in one factory but made up by increasing production at another, which happened to have spare capacity, the loss to the nation may be quite small, being just the difference in production

103

cost between the two factories. Thus the extent of indirect losses varies widely. The FHRC distinguishes between industrial activities according to their degrees of specialisation, of concentration and of difficulty of repair. In some circumstances the indirect damage-loss can be considerably greater than the direct cost. Other types of indirect loss considered in the FHRC methodology include disruption of services and communication, extra heating costs, as well as police, fire brigade and ambulance effort. Special care is needed to avoid double-counting. The figures given in the FHRC manuals are derived from a number of careful analyses of actual floods, with consideration of linkages and effects on competing factories, etc., all reduced to terms of unit area and flooding depth and duration. The manuals emphasise the need for users of these tables to understand the principles behind the valuations. Other countries may have their equivalent manuals and data sets.

Alongside the distinction between direct and indirect damage-loss, the manuals and other publications distinguish between tangible and intangible elements, the latter being those that cannot be expressed in numerical economic terms. This is not a hard and fast distinction, nor a particularly useful one, as with more effort and more research some intangibles can eventually be brought within the valued category. The disruption, stress and health effects of flooding are not easy to quantify, but surveys and interviews can give an indication. For some sorts of flooding (e.g. sudden floods without warning) and for particularly vulnerable people, the intangible component, however defined, may be larger than the quantified part of the flood's consequences; in such situations a numerical benefit estimate is likely to be a lower-bound one.

Although the British FHRC methodology does cover agricultural losses, there are other guidelines for these, notably, for the UK, Flood and Coastal Erosion Risk Management Appraisal Guidance (FCERM-AG).[23] Like the *Yellow Manual* of 1992, it covers salt-water flooding and non-flooding aspects of coastal work, such as reduction of beach erosion. It covers environmental effects, recreational values and non-use values. It employs most of the methods discussed in Appendix C of this book.

An instructive case study is described by Turner *et al.* (Coker and Richards, 1992, Ch. 5). The project was to strengthen the coastal defences near Aldeburgh in eastern England, where the likely breaching of old defences threatened a large area of land, all designated as 'of outstanding natural beauty', that included an 'environmentally sensitive area', a 'site of special scientific interest', and a 'national nature reserve', all these being official designations. The following valuations were undertaken:

(a) Agricultural losses avoided by preventing flooding of low-lying fields (lost production was estimated for irrigated and for non-irrigated land under two different degrees of flooding).
(b) Property damage avoided (hydrological and engineering studies defined the extent of flooding of buildings in terms of frequency, depth and duration, and national standard guidelines were used to put a value on the losses).
(c) Environmental damage avoided (a mixture of methods was used, including the market price for some of the land involved and the expected cost of local flood protection for other parts).
(d) Heritage and recreational losses avoided. (The loss of a historical monument, the best-preserved of 13 remaining Martello Towers, was valued as the cost of buying and restoring one of the other 12; loss of sailing clubs and boat yards was valued at the cost of replacing them with new facilities in the same estuary. This was an economic benefit because other existing facilities were fully booked; if that had not been the case the analysts would have omitted this item on the grounds that the sailors could have relocated to other clubs without loss of amenity, and the financial loss to some club operators would have been balanced by gains to others, thus becoming mere transfers from the national point of view. Other recreational benefits were valued by a carefully designed contingent valuation survey of actual visitors, supplemented by travel cost information for those coming from far enough away to be able to consider alternative sites.)

This example shows how complex a thorough CBA of a project with major environmental implications can become, and how a variety of different methods can be used. The analysts used links to real market prices wherever they could, and only fell back on contingent valuation for benefits that could not be tackled any other way. They then took special care to avoid the biases to which contingent valuation is known to be liable. To a small extent they also used the lower-bound approach, in that their contingent valuation survey only considered actual users of the recreational facilities, thus excluding the values, such as bequest and existence values, that other people might place on the facilities, although they did not use them at the time.[24]

If a flood alleviation CBA is done in a country that does not have standardised methods and unit costs of this kind, the analysts will have to collect the data directly as best they can with the resources and time they have. The analysis may have to be simplified to match these constraints.

4.6.3   *Recreation, landscape, sustainability*

Recreational benefits have been discussed above in connection with coastal improvements, because recreation is one of the main uses of coastal areas. In general, recreational benefits can be valued, although not with high precision, by means of many of the methods described in Appendix C, especially contingent valuation and the various revealed-preference methods described in Sections C.4 and C.5.[25]

To some extent an analyst can use unit benefit estimates from elsewhere (benefit transfer). A very rough indication of the levels of benefits in British conditions is given by the following list (pounds sterling at 1990 prices, by travel cost and contingent valuation methods):[26]

- Day visit to major recreation area (Lake District) or a beach: £5 to £9 per person per visit.
- Boating on a reservoir: £2 to £8 per person per day.
- Enjoyment of a canal: £0.2 to £0.6 per person per visit or per day.
- Extra enjoyment resulting from river quality improvement: £0.5 per person per visit.
- Annual value of river quality improvements: £15 to £20 per household (either living nearby or visiting).

The treatment of environmental values in CBA, as in planning and decision-making more generally, is largely based on the concept of sustainable development (improving the quality of human life while living within the carrying capacity of the supporting ecosystems). This is an example of the general need to optimise something subject to constraints, which is also the basis of linear and dynamic programming. In particular instances it is important to decide whether or not an environmental asset is substitutable: Can the loss of a butterfly habitat in Birnam Wood, which would be a consequence of Option 2b for the Dunsinane Bypass, be countered by the creation and preservation of an equivalent habitat elsewhere, or not? The concept of substitution of natural capital allows the loss of one environmental asset or piece of landscape to be balanced against the improvement or creation of another, whether as part of the project under consideration or as a *shadow project* used for costing. Where landscape or environmental goods are to be preserved in any case, a project has to be designed around such absolute constraints; the project may then become more costly than it would have been without the environmental constraints, but the environmental goods do not need to be valued for the project's CBA.[27]

Readers needing further guidance on valuing recreational benefits should use the references given at the end of Appendix C.

4.6.4 The lower-bound approach

It was mentioned in Section 4.6.1 that the analyst can sometimes simplify an analysis by evading some issues altogether, through judicious definition of the with-project and without-project situations. For example, a project was designed to rebuild and then protect a 20-km stretch of beach that was being eroded by sea waves, leaving only a narrow, evil-smelling clay-like foreshore.[28] The beach was used for bathing and recreation by roughly 20 000 people on a typical summer Sunday, and by lesser numbers of people on other days. Behind the beach was a surfaced road along the top of a low coastal ridge, and behind that was a mangrove swamp and an informal housing area that was, because of crowding on higher land, creeping ever lower into the landward fringe of the swamp. In one corner of the swamp was a sensitive ecological conservation area with rare butterflies. During ocean storms the waves were more and more frequently breaking on the road, which was crudely protected with rip-rap but was frequently damaged by the energy of the waves. The rip-rap and road surface were repeatedly patched, at considerable expense, between storms.

The technical judgement was that the energy of incoming ocean waves had in the past been dissipated to a considerable degree by the gradual slope of the beach, but that the loss of sand reduced this effect and left the road and its ridge exposed to much more wave energy than in the natural state (the natural littoral drift that should have supplied more sand was interrupted by harbour works). At the time of the analysis the damage to the road was becoming frequent and serious, even in unexceptional ocean storms, and it was estimated that without any action there would be a significant probability that an exceptional storm would break through the coastal ridge altogether and, by erosion and the influx of salt water, largely destroy the mangrove swamp and conservation area and endanger the residents of the swamp fringe. The project works comprised groynes and sand replenishment to restore the beach, both for bathing and for dissipation of wave energy.

The purpose of this particular CBA was only to justify the implementation of the project in economic terms, not to guide its design. Less than two weeks of an analyst's time was allocated to the CBA. The first step was to identify the project's costs and benefits. The costs consisted only of the beach replenishment plus the construction and subsequent maintenance of the groynes. The benefits were many and complex, and included: the continuing enjoyment of the beach by large numbers of people (including tourists); the protection of the mangrove swamp, conservation area and housing from sea water flooding; the reduced frequency of road interruption; and the reduced frequency and extent of repair work on the road surface and rip-rap.

All these benefits could have been assigned economic values, at least approximately, by some of the methods described in Appendix C. This would have been very time-consuming and expensive, and the results would probably have attracted controversy. Instead, the analyst and the project designers set out to find a way of justifying the project economically without the need to value all the benefits. This was done by judicious definition of the without-project situation. Instead of assuming that, in the absence of the beach replacement project, things would deteriorate until the ridge and road were breached and the hinterland flooded, it was assumed that the road and its rip-rap protection would have to be made the principal coastal defence element. (Strictly speaking, this assumed the hinterland was worth protecting, without proof or analysis; in practice this was a reasonable assumption in an analysis that involved many approximations anyway.) Accordingly, an estimate was made of the cost of repeatedly repairing the rip-rap and road in the future, without beach reinstatement and thus with increasingly frequent and extensive storm damage. This was not particularly easy to estimate, but still a great deal easier and more confidence-inspiring than an estimate of the economic value of the loss of environmental benefits, houses and possibly lives in the hinterland. Estimates were made both for the without-project situation (frequent major repairs) and for the with-project situation (modest routine maintenance, some repairs after the largest marine storms), and so when the incremental costs were calculated (with-project minus without-project) the result for the road was a negative cost. In the analysis, the costs of repairing and maintaining the road and rip-rap were placed in the costs part of the spreadsheet, as a negative incremental cost alongside the positive capital costs. They could have been shown as a positive incremental benefit, but it was neater to keep all the engineering elements in the same part of the spreadsheet, and, as no B/C ratio was calculated, it made no difference to any indicator.

This pragmatic definition of the without-project situation removed the need to put an economic value on many of the environmental impacts of the project, leaving the peoples' continued enjoyment of the beach as the main remaining benefit. For this a *travel cost* approach was considered. As explained in Appendix C, this involved finding how much money and time people spent in order to travel to the beach, and using this as an indicator of the value they placed on the enjoyment of the beach. It turned out that a few of the estimated 20 000 people using the beach on a summer Sunday were rich people from the far side of the city, coming by car and spending some tens of US dollars on the round trip. Some very poor people came on foot at no monetary cost at all, some slightly wealthier ones had bicycles, but the majority came a few kilometres by bus or truck, paying one US dollar or more. The range of travel costs was thus much wider than is usually the case with the travel cost method, and it was not felt that the method could be rigorously applied in this case, as the people's willingness-to-pay was so drastically affected by their ability to pay. (Although the theory of deducing willingness-to-pay from consumer surplus allows for a range of willingness-to-pay, it did not seem right to the analyst to assume that one bather on the beach derived a hundred or a thousand times more value from the day out than the poorer person on the next towel.) As a starting point, loosely guided by the travel cost considerations, an arbitrary estimate of one US dollar was taken as the average economic value of one person's day on the beach.

The rest of the analysis was performed in the usual way but rapidly and approximately, with shadow pricing of unskilled labour (shadow price factor (SPF) of 0.4 to 0.6 according to type and time of year) and of imported machinery and services (SPF = 1.2). In view of the uncertainty about the estimates, and not having the resources for a fuller analysis, the analyst took advantage of the fact that the CBA only had to justify the project, not guide its design: he made it a lower-bound analysis. (A lower-bound analysis is one aimed at finding the lowest reasonable estimate of a parameter or indicator, in this case the EIRR, that needs to exceed some threshold; for some contributing parameters, such as the costs in a normal CBA, it is, of course, the upper-bound value that is needed.) This means that the benefit estimates did not claim to be comprehensive, but to say of each aspect 'the benefit from this aspect of the project is at least X, perhaps more'. The analysis was calculated using a set of reasonable base-case assumptions on these lines, and produced an economic IRR of 19%, which was well above the decision-makers' cut-off value of 10%.

The preliminary conclusion was thus that the project was justified. Because of the hasty and rough nature of the analysis, the robustness of this conclusion was tested using a number of fairly drastic sensitivity tests, which produced the following useful conclusions:

- with the base-case assumptions for all other estimates, any assumed average beach-enjoyment value of more than US$0.20 would justify the project (i.e. this parameter, whose base-case value was US$1.00, had a switching value of only US$0.20)
- the estimates for road repair costs were not crucial (wide variations made little difference to the EIRR)
- omission of a hypothetical tourism benefit made little difference (the tourism element was included in the base case for the sake of completeness, but its forecasting was debatable because the area was plagued by civil war at the time, and the prospects for tourism did not look convincing; as it turned out, the war ended shortly afterwards, but an analysis that depended on forecasting that event would not have carried much weight at the time of the analysis).

It was thus possible to argue that the lower bound of reasonable estimates of the IRR was about 15%, so the project was economically justifiable. In effect, the beach enjoyment alone would justify the project, if the value of US$1.00 per person-day were accepted. The advantage of the lower-bound approach was that the admittedly extreme imprecision of most elements of the analysis did not matter. Although repeated repair of the road was not necessarily the best technical alternative to beach reinstatement as a means of protecting the hinterland from marine storms, it was a credible one. If its costs (claimed as benefits in the project's CBA) were perhaps a slight overestimate of the cheapest alternative's costs, it could always be argued that the benefits were being understated anyway, because the beach reinstatement would provide better protection against extreme storms than the assumed alternative.

A lower-bound approach is only possible in a CBA aimed at guiding a yes/no decision, and only works when the lower-bound indicator does turn out to be clearly above the critical or cut-off value. In a yes/no case it is often worthwhile to start by trying a lower-bound approach; if it produces a clear-cut result then time and effort can be saved by omitting a full analysis, while if it does not then a more careful analysis will still be needed, but will be guided by the initial lower-bound analysis and its sensitivity tests.

4.7   Transport

The use of CBA for projects, programmes and policy changes in the transport field is complicated by the wide variety of benefits and other impacts of any transport initiative, and the fact that some of these are much easier to put a value on than others. Many of the benefits of improvements to transport infrastructure are cost reductions, and the list of relevant benefits includes:

(a) User benefits (those perceived and enjoyed by people who use a road or other transport link):
  • reduced vehicle operating costs
  • time and effort savings
  • reduced accidents (user's perspective – pain, grief, lost earnings).
(b) Non-user benefits (those enjoyed by the rest of the nation or relevant population):
  • reduced accidents (community perspective – ambulances, police, hospitals, aftercare)
  • stimulation of economic activity (can be a negligible or negative benefit)
  • environmental changes (for better or worse) affecting people near a new link or the existing ones from which it diverts traffic – noise, air pollution
  • wider environmental effects such as regional or global air pollution effects
  • loss or enhancement of environmental and recreational facilities, or improved access to them.

The costs of a transport initiative can fall into similar categories, as well as the straightforward construction costs of a new bridge, port, railway or road; congestion and accidents are often made worse during construction and maintenance operations. Thus the valuing of costs is as complex as that of benefits, and the two are usually handled together (often just called 'impacts'), taking the with-project/without-project differences in each category. Because the distinction between costs and benefits is so arbitrary in this field, B/C ratios are seldom appropriate and NPV is the main indicator.

The physical extent of the impacts of a transport initiative are estimated by technical means such as traffic modelling, environmental studies and *dose–response relationships*, all of which are complex techniques in themselves. For CBA the analyst then proceeds to place a financial or economic value on each impact, and at this stage bias can

111

occur because both analysts and decision-makers may choose to ignore or play down some impacts just because they are difficult to value. Most of the methods mentioned in Appendix C are relevant, but many are also difficult to use, either because of the inherent difficulty of valuing things such as avoidance of loss of life, or because of vast data requirements. Pragmatic compromises are, of course, reached, often involving the exclusion of some impacts from the CBA. If these impacts are fully described, quantified, reported and used alongside the CBA as decision guides, little is lost. In practice, however, the residual narrow-based CBA often tends to dominate the decision process, with the result that the use of CBA in the transport field is often controversial.

The limitation of transport CBA within defined boundaries is illustrated by the British experience since the early 1970s. In the UK, the economic analysis of road schemes has since that time been codified and standardised in great detail in a computer program called COBA and its user manual, both of which have been periodically updated over the years. COBA 9 was superseded in 1996 by a revised version, parts of which are called COBA 10. This version is described in a revised manual, referred to in this book as COBA 1996, which uses 1994 prices. COBA 11 was issued in 2001 and revised in 2006, and is available online. COBA incorporates standard unit costs for travel time, vehicle operating costs and accident costs affecting road users (including pedestrians); the benefits of a road scheme are regarded as consisting mainly of reducing these costs, and so the important step is to define a without-project situation and one or more alternative with-project situations (called the 'do-minimum' and 'do-something' options, in COBA terminology). COBA also provides for construction costs, land and property costs, traffic delays during construction, and future costs, including maintenance work and the cost of any traffic delay that it causes. In many parts of the methodology there are default values built into the computer program, which the user can replace with case-specific estimates if desired. For instance, the default accident rate for a road network link (without junctions) of road type 'older D2 roads', with a speed limit of 40 mph, is 0.266 personal injury accidents per million vehicle-kilometres: this is given in a table alongside figures for 14 other road types and various speed limits, and other tables split casualties into fatal, serious and slight. All these terms are precisely defined, and more tables cover junctions of 96 defined types. The standard values of the cost per casualty were, in 1994 prices, £794 090 for a fatal casualty, £89 380 for a serious casualty and £6920 for a slight casualty.

One of the more difficult tasks in preparing a transport CBA is forecasting traffic flows. In computing benefits, the basic COBA program applies only reassignment, i.e. traffic travelling between the same origin and destination at the same time of day in the with-project and without-project situations, but by a different route; this is the *fixed trip matrix* assumption. If needed, other models and programs related to COBA can bring in *variable trip matrix* evaluation, i.e. generated traffic and road users who change the timing, origin or destination of their journeys, or their transport mode.

These selected features and examples give an indication of the extent of detail and refinement of the COBA system, which encapsulates in computer programs and manuals many years of research and the analysis of vast quantities of data on roads and traffic. So an analyst faced with the need for a CBA of a road scheme in the UK can buy the COBA system, estimate the effects of the scheme by traffic modelling, and put the scheme-specific data into the computer program, using COBA's default factors and values except where there is a special reason to use another estimate. All the aggregating, subtracting and discounting will be done automatically by the program. This is obviously very convenient, saving much effort on the part of people analysing individual schemes, and also ensuring a high degree of uniformity in the way competing road schemes are analysed. The danger of course is that COBA may be treated like a *black box*, a machine which requires a minimum of scheme-specific data on the input side and produces a ready-to-serve economic indicator on the output side, without the analyst having to think about all the myriad assumptions and factors and unit values that are built into the box. To counteract this danger, the COBA manual takes care to explain many of the principles involved and to state all the numerical assumptions and default values; it goes out of its way to be transparent. Like any complicated analytical tool, it can be misused by an irresponsible analyst who wants a black box, but it can also be responsibly used.

COBA is deliberately restricted to valuations that can be derived from willingness-to-pay measures, and this is both a strength (in minimising subjective bias) and a weakness in that it fails to take adequate account of many non-marketed impacts of roads, especially environmental impacts. A further criticism is that its application to road schemes, one by one, fails to take account of an adequately wide range of alternatives. In particular, the definition of the without-project situation as being a do-minimum policy on one (usually small) piece of the road network fails to take account of the possibility that the

113

scheme's objectives might be better met by other transport modes, such as rail. Decisions would be better guided if the thoroughness of COBA were applied to wider ranges of impacts and alternatives.[29]

There are other guidelines in other places, and a World Bank publication (Adler, 1987) covers a wide range of transport-related decisions, including the choice between a bridge and a ferry, the justification of a highway maintenance programme, the widening of a road, the electrification or dieselisation of a rail link, construction and modifications of ports, closing down of a rail service, and construction of a pipeline and an airport. Questions of timing are explicitly considered. The range of impacts considered is, however, quite limited: after dealing with reduced operating costs, stimulation of economic development, savings of time (both for people and for freight), and accident reduction, Adler lumps several others as secondary benefits that are mostly of an economic nature. Most non-user costs and benefits are ignored: neither the steam/electric/diesel comparison for a railway nor the air/road comparison make any mention of noise or pollution. The resulting simplistic analysis may be sufficient for some decisions, but leaves itself open to criticism and can lead to quite serious adverse consequences.

Although the loss of environmental assets in the course of building a road, railway or airport can, in principle, be estimated by means of one of the techniques mentioned in Appendix C, this is difficult in practice. In one case in northern England the choice of route for a bypass road depended on the value placed on preserving a wood.[30] A contingent valuation survey was conducted, using both willingness-to-pay and willingness-to-accept-compensation approaches. A thousand questionnaires were sent out to local residents, but only about a third were returned. Many of those were protest responses, generally indicating that protesting respondents did not accept the use of any finite amount of money to put a value on the preservation of the wood. Quantifiable responses, from the remaining small fraction of the people asked, ranged widely. The results were inconclusive, as the indicated decision between the routes depended on what population was considered to be relevant and represented by the respondents, whether the analysis excluded one outlying response that fell more than 10 standard deviations from the mean, and which of the two surveys was used (the willingness-to-accept one gave an estimate of value more than 10 000 times bigger than the willingness-to-pay one). In this case at least, contingent valuation did not help with decision-making.

In less developed countries many of the techniques developed for modifications to already dense road networks are not applicable, and the main need is to choose and justify a road and pavement design, with associated maintenance intentions, for a particular link. Since the 1970s, the World Bank and others have developed computer-based models, such as HDM-III, which incorporate much research on the linkage between road design, maintenance and vehicle operating costs. By adding exogenous costs and benefits of other sorts, which must be estimated outside the model, the same program can perform the arithmetic of CBA internally. Ultimately, the use of models is merely a convenient way of using the accumulated and distilled experience and research of other places and times, despite the heavy data needs: it often amounts to a sort of semi-automated benefit transfer. A conscious effort is then needed to maintain transparency and avoid using unjustified assumptions just because they happen to be built into the available model.

For further discussion and specific advice about transport CBA, for both projects and policies, the reader is referred to Adler (1987) and Layard and Glaister (1994, Chs 7, 12, 13 and 14). The treatment of risks, accidents and loss of life are discussed in Chapters 8 and 9 of Layard and Glaister (1994), and in Jones-Lee (1989). An example concerning road benefits in a developing country is given by Curry and Weiss (1993, pp. 167–171). Analysis of large transport projects is treated in Priemus *et al.* (2008), and further details are given in Brent (2009). The World Bank handbook (Belli *et al.*, 2001, Ch. 10) presents the principles, but with few examples and minimal attention to environmental aspects.

4.8 Energy

A power station or hydroelectric project produces an easily measurable output, electrical energy, but it is one whose value (in either economic or financial terms) varies according to the time of day, the season, the availability and characteristics of other energy sources, and the energy demand. Reliable peak energy is much more valuable than base-load energy or unreliable (non-firm) energy.

For financial analysis of a single project one can estimate the future tariffs (usually varying by time of day, etc.) and hence calculate the financial benefits directly, as the takings of the operating agency.

Estimating economic benefits is slightly more difficult. In principle, one could proceed (as is usually done for water supply) to estimate, by means of one of the methods described in Appendix C, willingness-to-pay, but distinguishing between peak and base and mid-load

115

and non-firm energy by such methods would be quite impractical. Depending on the decision to be guided, it may be sufficient to assume that future demand (according to whatever forecast the analyst chooses to use) will be met, and then look for the cheapest way to do that. This means studying the whole national or regional grid in various hypothetical future configurations, with and without the project under consideration and likely competitors: at grid level it is a least-cost analysis. One can then, if one wishes to talk of benefits from a single project, take these as being equal to the cost of the next cheapest alternative.

It is also possible, when the power project is small relative to the grid, to use this alternative-source logic without analysing the whole grid. For instance, for the economic analysis of a proposed hydropower project, one can define the without-project situation as the construction of a thermal station instead. Although in practice thermal stations would be fitted into a national grid rather differently from hydropower ones, it is convenient to define the without-project situation as the construction and operation of a hypothetical thermal station (or set of thermal stations) that produces exactly the same electrical energy as the proposed hydropower one, hour by hour and year by year over its whole lifetime. This makes the benefits identical for the two situations and reduces the whole analysis to a least-cost one, so that there is no need to put an economic value on the electrical energy by any direct method. Another way to talk about the same analysis is to refer to the hypothetical thermal system's costs as the benefits of the proposed hydropower system. The fictional alternative station must be the cheapest credible alternative, and thermal power may not always fulfil that condition; in any case if the proposed power plant is very large relative to the grid, one should not take this short cut, but analyse the whole grid, as in the Sila example described in Section 2.1 and Chapter 5.

The sensitivity tests in an energy sector analysis (financial or economic) should include variations in the future real cost of fossil fuels, as the financial world market price is politically volatile and the economic cost of non-traded fuel is difficult to estimate and forecast (both depend heavily on estimates of remaining reserves, which are largely unknown). Fossil fuels usually feature somewhere in the CBA for any power plant, either as inputs to a proposed thermal plant or as an element in the costs of a hypothetical alternative to a proposed plant of any other sort. It is wise to include not only variations in the near-future fuel cost, but also, independently, variations in its far-future trend (e.g. by including an escalation rate in the CBA). The

116

economic valuation of fuel may need to include a depletion premium, as described in Section 2.6.4.

Energy demand forecasts are also notoriously uncertain and need to be covered by wide-ranging sensitivity tests. Further information on CBA for hydropower projects is given in Merrett (1997).

4.9   Urban and regional planning

CBA can be used to guide some decisions in the field of *urban* and regional *planning*, but there are limitations. The narrower the scope of the decision, in terms of sectors and activities or of geographical area, the more likely is CBA to be useful. Applications of CBA to a particular investment or policy are covered by the relevant parts of this book, and by the more specialised literature that it refers to. Applications to wider local or regional planning decisions are possible, but the CBA must be grounded in a thorough analysis and understanding of the economy of the area concerned and its interactions with the rest of the country. In some ways one needs to apply to the planning zone some of the ideas, such as border pricing, that are usually applied to a nation. In the terms of Chapter 1, it is a matter of appropriate definition of the group of people on whose behalf, or from whose point of view, the analysis is being conducted. Because of this need for examination of intra- and inter-zone linkages, CBA for urban and regional planning decisions is a task for specialised economists, and is beyond the scope of this book, although founded on the principles it sets out. Often CBA is only one of several decision guides, and not a major one, within a multicriterion matrix or analysis framework. The reader is referred to Schofield (1987, especially Chs 14 and 15) and Massam (1988) for the multicriterion context. A case study is given in James (1994). Relevant papers sometimes appear in the journal *Progress in Planning*.

Notes

1 The example is simplified from one given in Kuiper (1971, Ch. 14).
2 For example: ADB guidelines, Adler (1987), Chervel and Le Gall (1976, 1978), Dixon *et al.* (1994), Fabre and Yung (1985), FAO (1986), Gittinger (1982), Herz *et al.* (1991), Klümper (1995), Murray and Lopez (1994), ODA (1988), Prescott (1993), Psacharopoulos (1987), Psacharopoulos and Woodall (1985), Ray (1984), Squire and van der Tak (1975), UNIDO (1972), Ward and Deren (1991), World Bank (1993) and Belli *et al.* (2001).

3 For example: Abelson (1996), Brent (1990, 1998), Curry and Weiss (1993, 2000), James (1994), Kirkpatrick and Weiss (1996), Lee and Mills (1983), Mills and Lee (1993), van Pelt (1993, 1994) and Winpenny (1991). Schofield (1987) also has a chapter devoted to the particular aspects of CBA in developing countries.

4 ODA (1988) and Belli *et al.* (2001).

5 Discussion and guidance can be found in Pearce (1994), and a case study in James (1994).

6 This summary follows the exposition at the start of Torrance (1986), with some variation of terminology.

7 When the terminology is from the health service's point of view, these are sometimes called indirect costs.

8 This distinction between process indicators and output indicators is examined further in Belli *et al.* (2001, Ch. 9).

9 For a discussion see Brent (2003), World Bank (1993, especially Box 1.3), Belli *et al.* (2001, Ch. 9), Murray and Lopez (1994) or Wagstaff (1991).

10 Belli *et al.* (2001, p. 115).

11 Brent (2003) is a good overall text. Torrance (1986) gives a general review of the field and details of ways of quantifying the relative values of different states of health. Schofield (1987, Ch. 13) gives a brief review of the methods available, while Wagstaff (1991) gives a discussion of QALYs and the relationship between efficiency and equity objectives. Morrison and Gyldmark (1992) cover the use of contingent valuation. Prescott (1993) gives an example of the technicalities of choosing between different ways of targeting a health activity, and Prescott and Warford (Lee and Mills, 1983, Ch. 7) describe the application to less developed countries. The World Bank's *World Development Report 1993* provides a general discussion, definition and use of DALYs, and their application to less developed countries (Belli *et al.*, 2001, Ch. 9). Other papers in Lee and Mills (1983), Mills and Lee (1993), Culyer (1991), Murray and Lopez (1994) and Brent (2009) may be of interest for some applications.

12 A succinct account is in Klümper (1995).

13 This is the price elasticity of demand (source: Klümper, 1995). More data, and a distinction between long- and short-term elasticity, are given in Winpenny (1994, Ch. 4).

14 Projects supplying water where there was no piped supply before, and other sorts of water supply project, are covered by the practical discussion given in Whittington and Swarna (1994); methods covered include contingent valuation and hedonic pricing. See also Merrett (1997).

15 The analysis can be simplified, or the need for data reduced to some degree, by assuming a special type of earnings function (called Mincerian) that relates earnings to amount of education and to years of post-education experience, the relationship to years Y being a parabolic one ($aY - bY^2$, a and b being positive, so that the function peaks after $a/2b$ years). For

more discussion see Johnes (1993, Chs 2 and 3). Belli *et al.* (2001, p. 91) give an example showing the value of a university education in terms of the discounted differential earnings between highschool and university graduates; the result is some $380 000; this is a very narrow financial analysis from the student's point of view only.

16 The gender bias in estimates of private returns to education, and ways of correcting for it, are discussed in Herz *et al.* (1991).

17 Herz *et al.* (1991) quantify some of these but do not bring them into the CBA. The World Bank handbook gives an example for education of girls in Pakistan, showing a B/C ratio of 1.4 from future health and demographic effects alone (Belli *et al.*, 2001).

18 See Curtin 1996, on both taxes and indicators.

19 McMahon, writing in Psacharopoulos (1987), gives figures for the USA, Egypt, the Philippines and the UK. A particularly high return is noted for electrical technicians, intermediate ones for engineers, low ones for teachers, and a negative figure for ministers of religion. This may reflect the inadequacy of methods that rely mainly on earnings and thus miss part of the benefits flowing from education.

20 Penning-Rowsell and Chatterton (1977).

21 Parker *et al.* (1987).

22 The *Yellow Manual* was Penning-Rowsell *et al.* (1992). Numerous further publications by the FRHC, including the *Multi-Coloured Manual*, can be found at: http://www.fhrc.mdx.ac.uk.

23 The British government guidance on coastal erosion and flooding is in a March 2010 draft document called *Flood and Coastal Erosion Risk Management Appraisal Guidance* (FCERM-AG) available on-line, which supersedes a number of manuals and supplements produced from 1999 onwards under the title *Flood and Coastal Defence Project Appraisal Guidance* (FCDPAG).

24 Another example of a flood control CBA is given in James (1994).

25 A brief description of the application of such methods to recreational benefits is given in Schofield (1987, Ch. 11).

26 Taken from a National Rivers Authority publication (Postle, 1993, Table 3.2).

27 The distinction between constraints and prices for landscape and environmental goods is discussed in Bowers and Hopkinson (1994).

28 This is the same example that was mentioned in Section 2.1(c).

29 Some more detailed criticisms and suggestions are given in Bateman *et al.* (1993).

30 This is the Harrogate–Knaresborough Bypass, as described by Hanley and Spash (1993, Section 12.3).

5

Reporting and presentation

Humpty Dumpty began again 'Impenetrability! That's what I say!'[1]

It has been repeatedly emphasised in the preceding chapters, starting with the distinct roles of analyst and decision-maker in Section 1.6, that a CBA needs to be not only well carried out but also well communicated to those who are going to use it to guide a decision. The chief merits needed are clarity and transparency, even at the expense of making a report slightly longer than it might have been.

Any report must be written with an understanding of what sort or sorts of reader it is intended for. The language and terminology can then be adjusted to what the readers understand or already know. If, for example, the only significant group of readers of a CBA report is officials of one financing agency, it can probably be assumed that they have a thorough understanding of CBA and of that agency's preferred terminology, and the report can be written tersely by using that terminology without further explanation: jargon is a reasonable and concise way of communicating between people who share the specialised vocabulary. So the CBA section of a financing agency's staff appraisal report will often be quite short.

If, however, the report will be used by several diverse groups of people, some of whom do not share the analyst's specialist vocabulary, the style will have to be a more discursive one that uses technical terms sparingly and always explains each one, at least on its first appearance or in a glossary. Sometimes it may be wise to use two words for one concept, each aimed at a different sector of the diverse audience. To some extent one can save space by referring to other books or reports for definitions of terms, but only if one can be sure that the readers will have easy access to the work cited: to say 'the depletion premium is calculated as set out in Appendix 6 of the ADB Guidelines' is an excellent saving of time and words for those who have the guidelines, but a source of irritation and confusion to those who do not.

Cost–benefit analysis – A practical guide
ISBN: 978-0-7277-4134-9

The reporting of a full CBA requires not only a good deal of detail but also a filtering and summarising process. For example, the feasibility report for a large irrigation project included an annex on economics and finance that comprised 55 pages and was supported by a further 55 pages of tables in six appendices. All this was just one of 14 annexes in the six-volume feasibility report, and the 70-page main report summarised that annex in six pages. The final level of distillation was a separate executive summary, the six pages of which compressed the economic and financial analyses into the following summary (note the absence of abbreviations, even EIRR being written out in full for a wide and unspecialised audience):

> *The study includes a detailed economic analysis of the whole project package, including all infrastructure and services. With the moderate crop yields that are expected because of the nature of the soils, and prices forecast by the usual methods, the economic internal rate of return is only 2.1%. If the social and infrastructure costs are omitted this rises to 3.3%, and sensitivity tests show that under no credible set of circumstances is it likely to rise above about 4%. Financial analysis shows that the project would require a loan of the order of US$230 million at 1986 prices.*

This illustrates the need to present a CBA at several levels of detail, each a summary or distillation of the level below. Report readers who wanted to check or refer to the assumptions and base estimates could look at the 110 pages in the annex, including dozens of similar tables, where they will find, for instance, the assumed monthly cost of a health assistant at a project clinic, the cost of a 7-tonne tipping grain trailer, the seed rates and economic unit costs for wheat on state farms in year 1 and year 10, and the prices, local and foreign market prospects, and demand forecasts for all five groups of crops. Decision-makers looking at the feasibility study as a whole would read all the six pages in the main report, where they would find all the key assumptions, the base-case result, and a summary of all the sensitivity tests; they could then refer to a few places in the annex for details that they consider important or questionable. Senior decision-makers, for instance at ministerial or bank board level, would read the executive summary in full, browse through the main report, and ask their technical advisors for comments on the annex.

Appendix I presents a compressed version of a CBA report for an imaginary project, a composite of several real ones, which serves as an example of presentation as well as CBA techniques. As with the

121

real feasibility study just mentioned, all the tables come from one spreadsheet; with suitably designed tables embedded in it, this both performs the analysis and presents the results. In this chapter there follow a few examples that illustrate ways of presenting aspects of an analysis. All are based on the author's own experience, which results in a preponderance of water resources themes, but the presentation aspects are applicable to any field. The quotes and figures are modified and slightly simplified excerpts from actual reports on projects in four continents.

5.1   Definition of the project

The feasibility study for rehabilitation of three existing pumped irrigation schemes showed that the proposed works could only address some of the constraints that were currently keeping agricultural production down. To avoid claiming benefits that would not in fact follow from a decision to implement the project, the report was careful in the definition of the with-project and without-project situations. The pumps were driven by imported diesel oil, and the report said:

> *One major constraint on the three schemes' production in recent years has been the restricted availability of fuel oil nationally. This in turn has been mainly attributable to foreign exchange shortages, and its alleviation is obviously outside the scope of the rehabilitation project. To analyse a future situation not involving this constraint would be irrelevant to the real context of the project. We have therefore defined the project as an improvement in the way of using the same level of fuel oil supply. This hypothetical assumption is only for the purpose of economic analysis: the design of structures such as measuring weirs is aimed at the maximum development of the schemes.*

The report went on to define four situations: the present situation, the without-project situation, the with-project situation, and a future full development situation with the rehabilitation project and an adequate national supply of fuel oil. The project was then justified on the difference between the with-project and without-project situations only. As well as clearly excluding benefits due to something not consequent on the rehabilitation project, this presentation also made very plain how the without-project situation differed from the present situation. As it turned out the project (like many rehabilitation projects which, in effect, take advantage of large sunk costs) had very favourable indicators: a B/C ratio of 3.2 and an EIRR of 44%. So it was clearly worth implementing, even if the national fuel oil supply did not improve. It was implemented, but the country's foreign exchange position then became even worse and the fuel oil supply to the three rehabilitated schemes was problematic. In retrospect, the analysts should perhaps have done a sensitivity test with rehabilitation followed by a worsening fuel oil supply, although at the time it might have been regarded as unduly pessimistic. After a few years of inadequate supplies, the country found that it had oil reserves, and in time it became an oil exporter, so the inclusion of the adequate-oil scenario in the original analysis was not foolish after all.

5.2   Methodology and assumptions

The feasibility study CBA whose executive summary was quoted above opened its main report chapter on project evaluation with the following statement of numéraire, assumptions and key parameters:

> *The evaluation of the project has been carried out by an economic CBA using border prices. The methodology follows the relevant guidelines published by the . . . Government in 1981 (Ref . . .). It has been discussed with the responsible department (. . . , authors of the guidelines) and with the . . . and (the UN agency's) economists. The treatment of foreign exchange is that it is not itself shadow-priced, but all local currency elements are multiplied by a standard conversion factor of 0.75, which is equivalent to a 1.33 factor on foreign exchange. Unskilled labour is shadow-priced by applying a factor of 0.67 relative to other local costs. Details are given in annex . . .*

Note how the analysis is anchored from the start in accepted guidelines whose origin and authority is explicitly stated. It is mentioned that the methodology has been discussed with specialists from two of the report's separate audiences (the government and the UN agency), which further establishes the authority of the chosen methods and parameters, without denying their partially subjective nature. The choice of numéraire is unambiguously stated, but with a minimum of jargon, avoiding the word numéraire, and two key factors are given in this first paragraph of the chapter. Finally, the reader is told where to look for further details if they should be needed.

5.3   Statement of price basis

The following are modified excerpts from another feasibility study:

All economic costs and benefits are at constant 1992 price levels, the estimating datum being December 1992. The economic prices used for internationally traded commodities are based on long-term (1990) price projections expressed in 1992 constant prices. These prices were taken from World Bank commodity price forecasts of July 1992. In September 1991 the rupee was devalued by 10% and the premium incentive for mercantile transactions was removed. In December 1992 the currency was devalued again by about 10% to an exchange rate of Rs 21.70 to the US$. It is considered that the new exchange rate is close to the currency's real economic value, which is supported by the fact that the unofficial exchange rate differs little from the new official rate. This implies a standard conversion factor close to unity, and for this reason all prices, including those of non-traded goods and services, have been converted at the new official exchange rate. The rates for major items of work have been estimated from current local prices and have been broken down into the following main components: unskilled labour, skilled labour, cement, steel, equipment, transport, and contractor's profit. The breakdown has been based on the (client's) analysis of rates.

5.4   Probabilistic aspects of benefit estimates

The next modified excerpt is from a report on a drainage improvement scheme:

For the probability of flooding with the present drain and pump system, the period of 25 years is subjected to extreme value probability analysis. The two recorded floods are assigned annual probabilities of 0.022 and 0.062. An estimated probability function is drawn so as to pass between these two points and below a hypothetical third-ranked flood with negligible flooded area. The product of probabilities and areas flooded according to this curve (i.e. the area under the curve) indicates an expected value of area flooded of 393 ha/year. [The diagram was of the same form as Figure 4.1, page 101, in this book, but with the physical axis showing area flooded because in this case economic loss was assumed directly proportional to area.] *The proposed improved system will not completely eliminate the risk of flooding but will reduce it drastically – the area flooded by rainfall combinations of return period 50 years or more would be reduced by a factor of about 30. The corresponding probability curve shows an expected value of area flooded of 16 ha/year. The proposed works thus result in a 96% reduction of expected flooding. Half this reduction pertains to floods of*

*return period 10 to 50 years, and a third to rarer floods (over 50 years).
As shown in section … (sensitivity tests), reduced flooding accounts
for only about 10% of the benefits so there is no point in improving the
precision of this estimate.*

5.5   Treatment of sensitive demand forecasts

In the power station study already referred to in Section 2.1 under the
nickname Sila, the state electricity authority, the client for the study,
was politically committed to a demand forecast that the consultant
considered unrealistic and liable to detract from the credibility of the
study. The consultant recounted in detail the derivation of the ministry's forecast, stated why he considered it unrealistic in the short term
(due to technical limitations of the grid, among other things), and
produced a lower forecast, already suggested by another consultant.
The section on demand forecasts concluded:

> *For the planning of the Sila power project the consultant has accepted
> the ministry forecast. The proposed planning period is within the
> Ministry forecast's long-term period. The forecast is also the basis for
> Government planning policies. However, due to the high growth rates
> in the short/medium forecast, past experience and the short-term
> economic prospects for the country, the Consultant wishes to use a
> decreased ministry forecast as a planning alternative.*

The Ministry's forecast was tactfully called the 'basic forecast', and the
other one the decreased forecast. Then the base case was computed
with the Ministry's forecast, but the whole analysis (which was a complex
one involving development sequences for many power stations in the
national grid over several decades) was repeated for the decreased
forecast. The consultant knew that the international or bilateral agencies
that might be approached to finance the project would share his reservations about the ministry's forecast and would regard the sensitivity
test as the definitive decision guide, but with this dual presentation the
consultant was able to sit on the fence and satisfy both audiences.

5.6   Optimisation of timing

The same Sila report of 1980 provides an example of the use of the
maximum-NPV criterion to guide a decision on timing, with a graphical
presentation of results. Figure 5.1 reproduces the report's main presentation, a graph of Sila commissioning date against total discounted

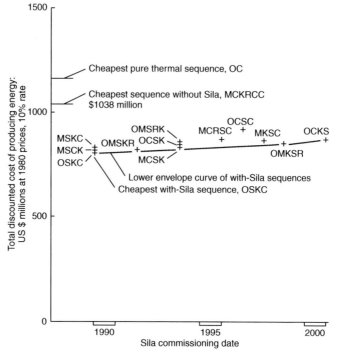

Figure 5.1 NPV comparison for project timing

economic cost for the development and operation of the country's power grid over the next 50 years. The letters, such as MSCK, represent sequences of power plants in order of construction, S being Sila itself, C being coal and O being oil for the thermal alternative, and M, K and R being other hydropower projects. The text said that:

> *the economic criteria show a slight preference for early construction of Sila, with the sequence OSKC (Sila 1990) as the most economic option by a small margin over OMSKR (1992) MSCK (1990) and MCSK (1994). These four have costs within 3% of each other.*

Note that the graph's cost axis has a true zero to avoid giving a false impression of the differences. The same figure shows graphically the cheapest option without Sila, namely MCKRCC with an NPV of US$1038 million. The report went on to deduce indications of Sila's economic merit, recommending one of the with-Sila sequences, MSCK, the NPV of which was US$819 million, and treating the lifetime costs of the cheapest without-Sila sequence, MCKRCC, as the benefits of MSCK. This gave a discounted cost ratio of $1038/819 = 1.29$. The report continued:

> *This can be interpreted as a benefit–cost ratio, if the costs of the next-best alternative, distributed through time, are regarded as benefits for the proposed course of action. With this concept the equalising rate has been calculated, i.e. the internal rate of return of the proposed option with the next-best option's costs as benefits. This is therefore the discount rate at which the two options have the same total discounted cost. The result is, in the base case, 16.0%.*

This report (which served multiple audiences) takes care to lead the reader gently to the slightly unfamiliar concept of next-best costs as project benefits, and hence of equalising rate as a particular sort of EIRR. The example also shows how, in complex cases, one must take care about finding the realistic next-best option: the simple concept of a thermal alternative, as described in Section 4.8 of this book, would have used the cheapest pure thermal sequence OC (also shown in Figure 5.1), which in this case would have quite seriously overstated the benefits of the preferred sequence.

5.7   Financial analysis

The Sila report also included an analysis of the project authority's financial incomings and outgoings if Sila were to be constructed. First it showed graphically how the before-financing FIRR, without any loans, depended on the price at which energy would be sold to users: at the country's then estimated economic energy cost of US$0.065/ kWh the FIRR was 11.9%. Then it analysed the authority's cash flow at various energy sale prices and loan interest rates. Figure 5.2 reproduces the graph for a 10% interest rate. This simple analysis gives an initial indication of the sort of loan package that would be required: one could go on to design a particular loan package and recalculate the cash flow under its terms. It is interesting to note that the financing requirement of more than a billion US$ effectively prevented this project from being implemented, despite the favourable economic indicator. In simple terms, Sila was too big for the country: its total capacity was 10 times the demand on the national grid at the time of the study, and even with division into multiple phases its first stage capacity would have doubled the grid's capacity. The initial investment for the dam was indivisible, and the dam could not sensibly be made smaller, for topographical reasons.

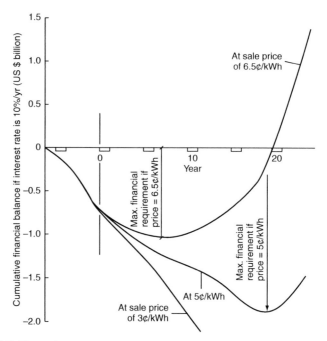

Figure 5.2 Financing requirement

5.8   Size optimisation and choice of indicator

Figure 5.3 reproduces the base-case result presentation in the report on
a size optimisation study. The proposed canal could have served
anything from 3500 to 198 000 ha of land, an unusually wide range of
possible project sizes. Eleven project variants were formulated, covering
the full range of possible sizes, and a separate CBA done for each, with
common assumptions. The graph of NPV against area served shows a
clear preference for variant 8, which had an EIRR of 22%. An accom-
panying table revealed that variants 1 to 6 inclusive all had higher
EIRRs, over 31% in the case of variant 2. Both in prior meetings and
in the report the consultant took care to explain to the client the need
to base the choice of optimum size on NPV, not EIRR (see Section
3.6), and the table showed the differential projects between the variants.
This demonstrated numerically that, with the exception of the local
maximum and minimum at variants 3 and 4, every step up the scale
gave a positive increment of NPV up to and including the step from
variant 7 to variant 8, while the further increase in size to variant 9 or
10, or on to variant 11, gave decreases of NPV and were therefore not
justifiable.

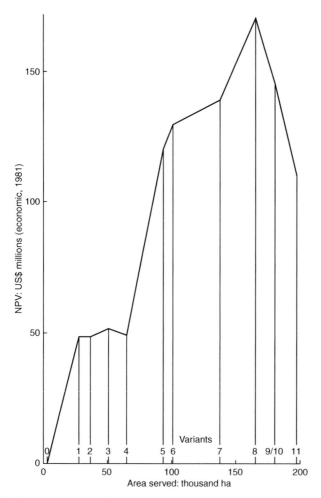

Figure 5.3 Size optimisation, base case

5.9 Graphical presentation of sensitivity test results

Figures 5.4 and 5.5 show two of the 11 sensitivity tests from the same report. Each is presented in terms of relative NPV, with the best variant's NPV as 100% and the base case represented by a dotted line for comparison with the full line for each sensitivity test. Figure 5.4 concerns the treatment of energy benefits, which were a side-issue in this study (for some variants the proposed canal would have affected an existing hydropower plant): it shows at a glance that the optimum remains clearly at variant 8.

129

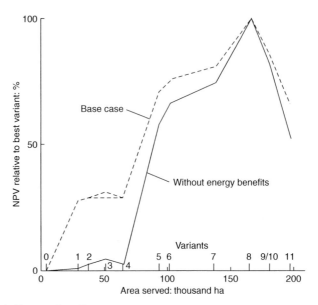

Figure 5.4 Energy–benefit sensitivity test on size optimisation

Figure 5.5 shows the sensitivity test for with-project agricultural production costs (see Section 4.2 for an explanation of how these can make a considerable difference to agricultural benefits if the net return is a small proportion of the gross return). Rather surprisingly, a mere 20% increase in these costs above the base-case estimate has a dramatic effect on the project size optimisation, with the optimum shifting to variant 6. The report gave all these results in tables too, but with 11 variants and 11 sensitivity tests the message would have been hard to discern from tables alone. The graphs, which were presented for eight of the tests in the standardised format of these two examples, were a very effective means of communication, not least because they focused attention on the relative NPVs and not on the actual numbers, which in many of the tests were necessarily different from the base case for all variants.

Both these sensitivity tests related to specific components of the analysis, and show how meaningful such tests can be. Although one of the 11 tests in this analysis was just a 20% reduction in all prices, most of them concerned specific factors whose sensitivity was of interest. They included, besides the two depicted here, tests on the rice and meat producer prices, crop yields, the lining of the canal, a delay in the benefits (which had a marked effect) and the mode of operation of an existing basin transfer scheme.

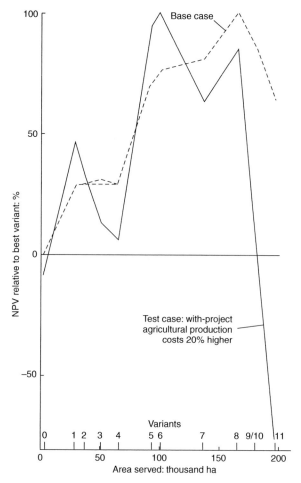

Figure 5.5 Agricultural cost sensitivity test from the same analysis

5.10   Sensitivity tests in text

The large feasibility report from which the first quote in this chapter was taken included the following verbal descriptions (a selected 10 out of the 12 tests reported) of sensitivity tests in the economics annex:

(a) *Social and infrastructure costs: the omission of these reduced initial economic costs by 31%. Recurrent costs were not reduced. The result was that the EIRR rose from 2.1 to 3.3%.*

(b) *Omission of provision for the existing semi-nomad population: the land occupied by the 4180 ha of settlement areas in the base case is practically all either unsuited to state farm crops or, in patches of better soils,*

131

too small and scattered to be economically irrigable. For this sensitivity test, therefore, a hypothetical scheme was analysed which irrigates only the state farm areas. This reduced the initial cost by only 9%, and the recurrent cost was also reduced by omitting the relevant elements, which also happened to be about 9% of the total. These changes only raised the EIRR by 0.35%.

(c) *Cropping pattern*: to test the importance of this, the growing of wheat and maize was omitted, which lowered the EIRR by 1.8% to 0.3%.

(d) *Delay to benefits*: delaying the start of benefits (and of recurrent costs) had almost no effect on the EIRR, as would be expected at low discount rates.

(e) *Cotton yields*: raising the estimate of cotton yields by 30% (i.e. from 2.4 to 3.1 t/ha for the cotton–wheat system) raised the EIRR by 0.9% to 3.0%: this also gives an indication of the likely effect of any increase in cotton yields that may be achieved by improving management or developing new varieties.

(f) *All benefits*: 30% increase in net benefits would raise EIRR by 1.7% to 3.8%.

(g) *Labour*: doubling the economic cost of agricultural labour, which might be postulated in view of the local labour shortage despite good national availability, pushed the rate down by 0.5% to 1.6%.

(h) *Recurrent costs*: a 50% decrease in the engineering recurrent costs, from 4.2 to 2.1 million pesos (economic) per year, only raised the EIRR from 2.11% to 2.65%.

Finally, two further sensitivity tests took the form of switching values. The value of this calculation when applied to a project of very low EIRR may be doubted, but it does serve to indicate how far short of conventional rates the project is.

(i) *If the internal rate were to rise to 12% because of increased cotton and tobacco yields, those yields would have to rise by a factor of 3.5. This is, of course, outside the range of the crops' potential.*

(j) *If the rate were to rise to 12% due to reduced costs alone, the costs would have to drop to 20% of their estimated level.*
    These sensitivity tests show that under no credible combination of circumstances can the EIRR rise above about 4%.

These words were then commented on in the feasibility study's economics annex, and summarised with the aid of a table in the main report. The final sentence of this quote, '... *no credible combination of circumstances...*', appeared again not only in the main report but also, as already quoted, in the six-page executive summary of the whole feasibility study. Some points of interest in the presentation of the sensitivity tests are as follows:

- Most of them refer to specific features of the particular project, rather than being across-the-board changes such as a reduction in all benefits as in test (f).
- The features were ones that the report authors knew were under discussion by one or another of the multiple audiences as possible reasons why the CBA's negative outcome should be challenged, so many of these sensitivity tests were aimed at forestalling criticisms by showing that the possible arguments for reversing the recommendation had already been analysed.
- To avoid giving a false impression of precision, the EIRR and its changes are only quoted to one decimal place, except where the changes are very small in tests (b) and (h).
- As the base-case EIRR was only 2.1% per year, the probable decision was to reject the project: for this reason many of the tests were done in the direction of higher EIRR, i.e. the direction that would reverse the decision; if the EIRR had been favourable they would have been done in the other direction.

5.11   Summary

This chapter has given some general indications about the presentation of CBAs, and several examples from real World Bank reports are given in Belli *et al.* (2001). An example of a tabular presentation of sensitivity tests is given in Appendix I. The main principles of such presentation are:

- transparency
- that enough information should be given to enable a report reader to trace and check every step of the calculations
- that there must be distillations or summaries, perhaps at more than one level, for readers who want just the main features of the analysis, without the details.

Note

1 *Through the Looking-Glass*, Lewis Carroll (C.L. Dodgson), Macmillan 1871, Ch. 6.

6

Checklist

This chapter presents a pragmatic checklist of the aspects that may need attention in both carrying out and reporting a CBA. For any particular CBA many of these aspects may be inapplicable, but by using a checklist one can ensure that such aspects are omitted consciously and for good reasons, rather than because they have not been thought about. As well as serving to check for possible omissions from the analysis, the list can form the basis for planning the structure of a report on a CBA. The list follows the sequence that was introduced at the end of Chapter 1, and is presented as a series of questions that the analysts (or a reviewer) could put to themselves both when planning the analysis and when reporting it. Some of the questions are amplified by a few examples or cross-references to earlier chapters.

(a) Define the decisions that are to be guided.
- What precisely is the project or course of action (or the alternatives in the case of a choice between technical alternatives)? What are its objectives (if there are several, how will the CBA deal with that)? Is the with-project situation adequately defined and described? Is it technically, institutionally and politically feasible?
- What would really happen if the defined course of action is not followed? Are there several plausible without-project situations and, if so, how is one chosen for analysis purposes? Would a different choice make the analysis easier to carry out or to explain? Is the chosen one adequately defined and described?
- Are all the costs and benefits of the without-project situation identified?
- Are there separable project components? If so, would separating them lead to one or more better projects? Are bad components avoidably riding on good ones?
- Is the project's internal design, packaging and timing optimised, to a reasonable degree of precision, by the same criteria that will be used to judge its overall merit?

(b) Define the group of people whose point of view is to be applied.
  - Is this a financial, economic or social analysis? (There may be several interlocking CBAs of different types, in which case this whole checklist should be applied separately to each one.)
  - Is there a subgroup whose interests should receive special attention, such as the rural poor within the nation on whose behalf an economic (or social) CBA is being done?
(c) Define the methods, criteria and main parameters.
  - How is the numéraire defined, and in what currency is it to be expressed? (See Section 2.3 and Appendix B.)
  - What date is used as the basis for financial prices?
  - What discount rate is used, and why? Is it appropriate to this analysis? Are sensitivity tests needed at different discount rates?
  - What indicator(s) are used and why? Are they correct for the decisions? (See Section 3.6.)
  - What analysis period is used, and how does it relate to the probable physical or commercial lifespan of the main components? Do some components have shorter lifespans, and if so are they covered by replacement costs? Are there any large residual values or future costs remaining at the end of the analysis period, and how are they dealt with?
  - How are the boundaries drawn for analysis purposes? Could they be adjusted to make the analysis easier, more precise, or more transparent? (Examples are given in Section 4.6.4 and Chapter 5.)
  - Can and should a lower-bound approach be used? (See Section 4.6.4.)
  - If it is not a lower-bound analysis, does the analysis capture all the relevant costs and benefits?
  - How are the costs and benefits expressed in economic (or social) terms? If shadow-price factors are used, how are they arrived at and justified (see Section 2.5)? Are they precise enough for the purpose?
  - Are there any cost or benefit categories whose financial prices are expected to change in the future at a rate markedly different from general inflation? Or any whose economic value will change? In either case, is escalation allowed for in the computation – in the base case or as a sensitivity test?
  - Are there some effects that could be treated computationally either as positive benefits or as negative costs? If so, which arrangement is better?

(d) Calculate the benefits.
- Are all the incremental benefits identified, quantified and valued so that they correspond wholly and solely to the decision that the analysis is guiding? Have all relevant benefits of the without-project situation been considered?
- Do the project outputs replace other production within the country? If so, have the implications been thought through?
- Will the annual incremental benefits in the project's operation phase reach their normal level instantaneously when the project components are commissioned, or will there be a gradual build-up for one or more reasons (see Sections 3.3 and 4.2)? How can its speed be estimated?

(e) Calculate the costs.
- Are all the incremental costs identified, quantified and valued so that they correspond wholly and solely to the decision that the analysis is guiding? Have all relevant costs of the without-project situation been considered?
- Have sunk costs been identified and omitted from the analysis? Are they really irretrievably sunk or would they have some value in the without-project situation?
- Are all ongoing, recurrent or annual costs of the project's steady operation phase included (operation and maintenance, repairs, lubrication, energy inputs, training of replacement staff)?
- Are all periodic costs included (replacement of electrical, mechanical, or fast-wearing components, painting)?
- Are any large late costs properly included (decommissioning and long-term environmental effects of a nuclear power station)?

(f) The net benefit stream.
- Is it necessary to show the net benefits, year by year? (The cash flow; if costs and benefits are discounted separately, for instance in order to calculate a B/C ratio, calculation and presentation of the net benefits may be redundant.)

(g) Calculation of the indicators.
- Is the calculation format convenient for result presentation and sensitivity tests? Can a suitably designed spreadsheet be used for both calculation and reporting?

(h) Sensitivity tests (see Section 3.7).
- Are all relevant variables and sources of uncertainty covered?
- Are there any matters of special concern to some or all of the decision-makers, or other affected parties, which should be

included as sensitivity tests for purposes of communication or persuasion even if not strictly needed for analysis?
- Can the sensitivity tests be done in terms of physically meaningful changes instead of arbitrary percentage shifts?
- Would some switching values be helpful to decision-makers? (The World Bank, among others, now requires switching values for at least some variables.)
- Is there a case for varying two parameters at once?
- Is there a case for probabilistic risk analysis? (See Appendix A.)
- Do correlations or causal links between parameters need to be considered?

(i) Reporting.
- What aspects have been consciously excluded from the CBA and need to be emphasised alongside its results (environmental or social impacts, unquantifiable effects, institutional aspects)?
- Are the CBA results to be used in a multicriterion analysis alongside other aspects?
- Have correlations or potential double-counting between the CBA and other aspects been thoroughly thought through and transparently reported?
- Are the concepts of viability and desirability adequately distinguished? (See Sections 3.8 and 4.2.)
- Is the report aimed at multiple audiences with different perspectives, needing different information? If so, will it satisfy all the audiences?
- Will the report, or data sources clearly referred to in it, enable a reader or enquirer to check the whole chain of computation?
- Does the reporting and summarising of results constitute an analyst's usurpation of a decision-maker's role?

A particular CBA can be quite simple and its report quite brief, as in any one case many of the above questions may not be applicable, but it is still worthwhile to go through a checklist like this one so as to avoid inadvertently omitting something relevant.

This checklist concludes the six chapters and thus the presentation of the main aspects of CBA, generally aimed at non-economists. Appendices A to G, with their notes and those of the chapters, give details and many literature and internet references so that readers who need to can examine particular aspects more thoroughly. Appendices H to K contain examples, discounting tables and the references themselves.

Appendices

A   Uncertainty, probability and risk

B   The domestic pricing and foreign exchange numéraires

C   Ways of estimating economic prices

D   Distribution of costs and benefits between people

E   Choice of discount rate

F   Multicriterion decision analysis

G   The effects method

H   Model answers to readers' worked examples

I   Worked example

J   Discounting tables

K   Bibliography

A

Uncertainty, probability and risk

A.1 Basic concepts

Cost–benefit analysis (CBA) concerns future actions, and all quantitative statements about such actions and their consequences have a significant degree of uncertainty. Whenever there is more than one possible outcome of an action or investigation, there is uncertainty. Usually there is some information about the relative probability of each outcome, even if it is only subjective or rough. The words *risk* and *uncertainty* are sometimes used in special senses in the CBA context, as shown in Box A1, but the distinctions are generally blurred and matters of degree rather than hard and sharp.[1]

The important concept is that there are two sorts of statement about any outcome or parameter: a simple statement of its magnitude and a more complex statement about the probabilities of different outcomes. The latter sort of statement contains much more information than the former (information can be expensive), and it is also much more useful. Probability information converts unquantified uncertainty into quantified uncertainty (sometimes called risk):

$$\boxed{\text{Unquantified uncertainty}} + \boxed{\text{Probability information}}$$

$$= \boxed{\text{Quantified uncertainty}}$$

Probability estimates can be entirely subjective, as when an expert estimates the probability of a project being disrupted by civil war during the construction period as 5%, or an estimator guesses that the chances of the project costing more than 1.4 times his best-estimate is 15%. Other probability estimates can be, in varying degree, objective. This applies especially to the case of statistical uncertainty as defined in Box A1, where statistical techniques enable the past history of some parameter or phenomenon (maximum 3-hour rainfall in Aberdeen, size of earthquake at a certain site, peak wind speed at a bridge site) to

> ### Box A1 *Terminology about uncertainty*
>
> *Dictionary definitions*
> Uncertainty: matters that cannot be accurately known or predicted
> Risk: possibility of incurring misfortune or loss
>
> *Conventional CBA definitions*
> Uncertainty: probability unknown, undefined, or no consensus
> Risk: probability distribution known or estimated
> Risk analysis: the term is sometimes used in a wide sense to cover a matrix of different sorts of risk, and sometimes to indicate analysis of probability distributions
>
> *Types of uncertainty*
> Statistical uncertainty: capable of extensive repetition, often amenable to frequency analysis of past experience
> Non-statistical uncertainty: essentially unique, so probabilities must usually be subjectively estimated

indicate fairly precisely the degree of accuracy or uncertainty of the estimate.

A few statements are, by their nature exact, such as 'an inch is 25.4 millimetres' or 'four people live in this house'. Most numerical statements, however, have a degree of uncertainty, even if it is not stated: 'she is 37' normally means her age is 37.5 ± 0.5 years, and 'it's four o'clock' usually implies a few minutes leeway (or at least as many seconds as it takes to say the words). Relative precision and uncertainty are quantified by statistical *probability distributions*, which are described by *histograms*, *central measures* and *dispersion measures*. Details of these can be found in any good book on statistics (and many hydrology texts). Box A2 summarises the most important definitions and Figure A1 illustrates four distributions by showing their histograms.[2] The top two histograms are particular non-standard distributions, one continuous and the other discrete, but both asymmetrical; the lower two are mathematical standard shapes that are convenient for approximations because they can be described with much less information.

The *expected value* or arithmetic mean of a probability distribution is the most important parameter for CBA. A particular case is the calculation of average annual damage or *cost-of-risk*. Suppose a certain proposed bridge is at some small risk of being damaged or even washed away by a large flood in the river it spans. The designer

Box A2 *Terminology of probability distributions*

A *probability distribution* is a statement that the outcome of an event will have particular values, or fall within particular ranges, with stated probabilities. If the event is, by its nature, repeatable (and if the probability distribution is stationary), the probability distribution is the same as the frequency distribution when the event is repeated a very large number of times.

The distribution can be most fully described by a *histogram*, which is a diagram showing the probability for each possible outcome or range of outcomes. Numerically it can be partially described by *central measures* and *dispersion measures*. There are three commonly used central measures:

- the *mean* or *expected value* (the average if the event were repeated many times)
- the *mode* (fashionable) or most common value
- the *median*, such that there is a 50% chance of its being exceeded.

The usual measure of dispersion is the *standard deviation*: for most distributions about 68% of the probability lies in a range two standard deviations wide, centred on the mean, and about 94% lies in a similar range four standard deviations wide. The standard deviation is an absolute dispersion measure, and the corresponding relative one is the *coefficient of variation*, which is merely the standard deviation divided by the mean. There are further measures that can be used to define skew and asymmetry.

For example, suppose the depth of a rock surface below ground level in a certain area has the probability distribution shown in Figure A1(a), which as a grouped approximation is:

0–5 m	10%	or	0.10
5–10 m	15%	or	0.15
10–15 m	20%	or	0.20
15–20 m	45%	or	0.45
20–25 m	10%	or	0.10

This describes five outcomes, which at any one place are mutually exclusive because only one of them can hold, and the sum of their probabilities is 1.0. In this case the distribution is asymmetric (negatively skewed), and the mean, median and mode are not the same (they are in fact about 14.0, 15.6 and 17.5 m, respectively). The standard deviation is about 5.7 m and the coefficient of variation is 0.4. Despite the skew, the 68% and 94% rules of thumb are very close to the truth.

The above is a *continuous distribution*, the histogram for which is a smooth curve that could, for arithmetical analysis, be approximated by a stepped line defining a number of groups (in this example five). Figure A1(b) shows a *discrete distribution*, namely that of the number of children per household in Striterax: there is a 25% probability that a household chosen at random will have no children, a 15% probability of one child, 30% for two children, and 15%, 5%, 4%, 3%, 2% and 1% for three to eight children, respectively. The modal (fashionable) number is evidently two children, but the distribution has a second mode at no children (it is *bimodal*). This distribution is also very asymmetrical, but by chance its mean, median and mode all coincide at 2.0 children (showing that asymmetry does not necessarily separate the three central measures). Its standard deviation is about 1.8 children, and it can be seen that 60% of households are expected to have one, two or three children, falling within the two standard deviations range centred on the mean; exactly 94% fall within the four standard deviations range, so the standard deviation is not a bad measure of dispersion even for this odd-shaped distribution. The coefficient of variation is 0.9, more than twice that in Figure A1(a).

Figures A1(c) and A1(d) show two standard shapes that are often used as convenient approximations to real probability distributions: The former is the *normal* or *Gaussian distribution*, which is symmetrical and so its three central measures coincide, and the latter is a *triangular* distribution, in this case an asymmetrical one. These standard shapes require much less information to define them than non-standard ones. The normal distribution is defined by two numbers, its mean and standard deviation, and the triangular distribution is defined by three numbers, the two extremes and the mode (the mean or expected value is then the mean of these three numbers). A useful variation on the normal distribution is the *log–normal* one, where the logarithm of the variable is normally distributed, so the variable itself has a smooth but asymmetrical distribution.

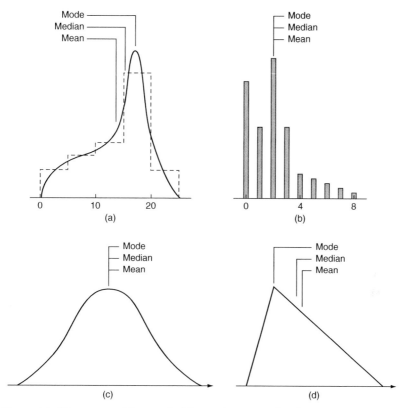

Figure A1 Histograms of four probability or frequency distributions

cannot economically eliminate the risk altogether, so the problem is to balance the benefit of increased robustness or flood resistance, in terms of lower expected damage, against the extra cost it would require, and thus to find an optimum bridge design. In order to use CBA to guide this decision, we need to estimate the annual cost that is equivalent to the future (and thus unknowable) damage that will be done by floods to the bridge, and to other infrastructure or people if such damage is influenced by the design of the bridge. This annual cost can then be included in a CBA for each year that the bridge will be in use, and can be discounted and balanced against other costs and benefits in the usual way.

There are two parts to the calculation of equivalent annual damage or cost-of-risk: the probability part and the sensitivity part. The probability part is, in the case of a flood hazard, hydrology. First, the wide range of possible flood peaks is divided into a practicable number of outcomes, each a slice of the probability spectrum. Together the list of outcomes must cover the whole range of probabilities, and they need to be

145

mutually exclusive, i.e. to have no overlaps. If these two conditions are met the probabilities will add up to 1.0. For this purpose, suppose the spectrum is divided into six slices at the 1.0, 0.05, 0.02, 0.01, 0.005, 0.001 and 0 probability levels, in terms of probability of being exceeded in any one hydrological year. These correspond to return periods of 1, 20, 50, 100, 200, 1000 and ∞ years, where the return period is the reciprocal of annual exceedence probability. By standard flood frequency analysis and extrapolation it is estimated that the flood peak discharges for these probability levels are 25, 740, 900, 1000, 1100 and 1350 m³/s, respectively, and of course an infinite discharge for the impossible flood (zero exceedence probability). (These are annual peak flows; to simplify the analysis the chance of having a second large flood in one and the same year, after the annual peak flood, is ignored, on the grounds that if it did happen the extra damage would not be large because the damage from the main flood would not yet have been repaired.) The results of these hydrological estimates are shown in the left-hand part of Table A1; the seven probability levels have divided the full probability spectrum into six slices.

The second part of an annual damage or cost-of-risk estimate is the sensitivity part, which quantifies the extent of damage for each outcome, in this case for each slice of the probability spectrum or flood range. For a bridge this is mainly an engineering task. The variation of damage extent with flood peak will not be a smooth function at all, in fact it is inherently likely to have thresholds and steps. Suppose

Table A1 Example of annual damage cost calculation for a bridge

Probabilities (flood hydrology)			Damage per flood event: £'000	Cost-of-risk (probability × damage): £'000/year
Peak flow: m³/s	Exceedence probability: per year	Slice probability: per year		
25	1			
		0.95	0.1	0.095
740	0.05			
		0.03	50	1.5
900	0.02			
		0.01	150	1.5
1000	0.01			
		0.005	1000	5
1100	0.005			
		0.004	1500	6
1350	0.001			
		0.001	2000	2
∞	0			
Total annual expected damage (cost-of-risk)				16

the bridge variant under study is designed to resist the 20-year flood (the one with 5% annual exceedence probability) without any damage worse than a little erosion of guide banks and some bent handrails. When the flood peak exceeds that threshold, however, the left approach embankment is expected to wash out, involving significant but not vast repair costs for the embankment itself and some consequential damage downstream. For floods bigger than the 50-year flood (2% probability), another threshold would be reached, and the right abutment is expected also to wash out, but the concrete structure would not suffer more than cosmetic damage until the 100-year flood, which is expected to undermine the central pier and cause a partial collapse of the deck. (The way the probability spectrum was sliced up in the hydrological estimate was deliberately adapted to these thresholds.) As a reasonable approximation it is assumed that each slice of the flood peak range can be represented by one flood size somewhere in the middle of the slice, and for each such flood the cost is estimated, including bridge repair, repair of any downstream damage, and appropriate allowance for any other damage or cost, including loss of life if it is expected with the more extreme floods. All these can be in financial or economic terms according to the sort of decision needed.

The fundamental relationship that enables us to bring the hydrological and engineering estimates together is the definition that, for any one outcome:

$$\boxed{\text{Cost-of-risk}} = \boxed{\text{Probability of flood}} \times \boxed{\text{Damage caused by flood}}$$

This is applied to each slice (outcome) separately, and because they are mutually exclusive outcomes the resulting costs can be added to give the total annual equivalent cost for this particular bridge design variant. People are sometimes uneasy about adding the cost figures for different sized floods in this way, as it is assumed that only one of them will happen in any one year. The computation is, however, only the finding of a weighted mean between vastly different possible outcomes, the weights being their relative probabilities.

Table A1 shows the numerical calculations for this bridge example.[3] One of the advantages of this way of setting out the calculation is that it shows which slices of the probability spectrum contribute most to the expected annual damage cost, and that information is valuable as a guide to the design of a better bridge variant. As the cost-of-risk for a slice is the product of probability and damage, some of the slices have small cost-of-risk because of small damage despite high probability (small floods), and some will have relatively small cost-of-risk because of small probability despite large damage (very large floods); the bulk of the expected annual damage or cost-of-risk is in the intermediate part of the probability spectrum. In this example the table shows that design changes to reduce the damage from floods of less than about 1000 m³/s would hardly be worth the trouble, while design refinements to reduce the damage from larger floods would make quite a difference. Rare floods in the range 1000–1350 m³/s contribute two-thirds of all the annual equivalent damage, so measures to reduce the damage they cause by half a million pounds would knock £4500 off the annual figure, and if this annual cost were being discounted at 6% over a life of 40 years (PV/RV factor of 15) the saving in damage would justify a cost increase of up to £65 000.

The summation of probabilities multiplied by damages can also be done graphically, and Figure A2 shows the graphical version of the same example. It plots damage per flood event against probability and flood peak discharge for the extreme twentieth of the probability spectrum, the rest being of no significance in this case. The damage and probability scales are linear, and so their product can be estimated by measuring the area under the graph; the rectangle shows how the area scale is determined. The dotted line corresponds to the crude lumping of the function into six slices, as in Table A1, while the full line shows roughly the real function, which has marked slope changes

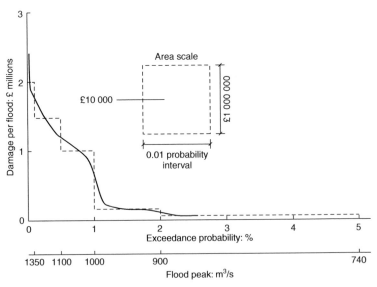

Figure A2 Damage/probability curve for the bridge example

near the 1%, 2% and 5% floods for the physical reasons explained above. The graphical version helps to give a feel for the relative importance of the different parts of the probability spectrum, and to emphasise how approximate the whole exercise is: to quote the result to more than two significant figures would be ridiculous. But this approximate method is much better than a guess: faced with this range of flood damage figures from £100 to £2 million, who could have guessed and justified a reasonable figure like £16 000 for the weighted mean?

Similar calculations can be done for the design or evaluation of anything that involves statistical probability, especially dam spillways and coastal protection works. By repeating the estimate for several different design options, the economic optimum can be found, to a reasonable degree of precision. This is, of course, the technically mutually exclusive case outlined in Section 3.6, and the relevant economic indicator is NPV, so the annual expected damage (cost-of-risk) is applied for every relevant year and discounted to find its impact on NPV, where it trades off against differences in construction cost. Similar calculations for with-project and without-project situations give estimates of incremental benefits of risk-reducing projects.

The same principle applies with non-statistical uncertainty, such as different outcomes or outturns for a project that depends on future demand which is hard to predict. For this situation, where 'risk' and 'damage' are not appropriate terms, the principle can be expressed as:

the expected outcome is the sum of each possible outturn multiplied by its probability.[4]

One major source of uncertainty is, of course, climate change. Although it is very difficult to estimate the probability of any of the more dire future scenarios, the fact that their damage would be vast makes the product 'damage × probability' very significant.

A.2   Attitudes to uncertainty

When the expected value of a physical or monetary parameter is calculated and then used directly in CBA as described above, the implication is that the decision-maker is *risk-neutral*. This term means that a certain shift in the parameter will always have the same importance to the decision-maker, whatever the starting point or direction of the shift. Suppose my total worldly wealth, mostly in the form of a house, amounts to £100 000, and I am considering some very uncertain business venture in which I might gain or lose £50 000. Would I regard gaining that sum

149

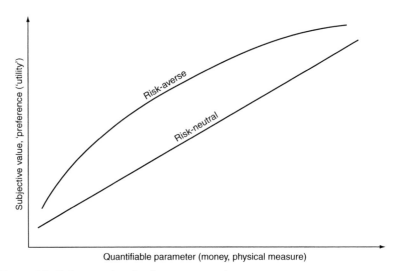

Figure A3 Risk-neutral and risk-averse attitudes

as having the same sort of significance as losing it? Probably not: gaining £50 000 would give a pleasant boost to my lifestyle but not change my situation drastically, while losing half my life savings would be cata-strophic, requiring me to sell the house for a start. So I would not regard the step from £100 000 to £150 000 as being as important as the numerically equal step from £50 000 to £100 000. This means that my preference or subjective value (or utility in the jargon) is not linearly related to money; I am not risk-neutral but *risk-averse.* Figure A3 shows graphically the risk-neutral and risk-averse attitudes.[5] The slope of the risk-averse person's function changes along the graph; by analogy one could say that the shadow price of the quantifiable parameter is changing. For small changes the slope varies very little along the change, so the assumption of a single conversion rate from parameter to value (a single slope or shadow price) is not far wrong. However, when drastic, wide-ranging changes are in question the chan-ging slope does matter. Drastic changes become relevant particularly in connection with environmental impacts, so it is in that context that risk aversion is most often discussed.

Most individuals are risk-averse, and governments are supposed to apply the collective (social) preferences of the individuals they repre-sent or govern. There is, however, an argument that governments should be risk-neutral.[6] The argument is that a government represents a very large number of people, and risk can be pooled and spread so that even a drastic gain or loss at national level is quite small for each individual, so the shift on anyone's *utility curve* is small enough for the slope to be effectively constant.

The idea is related to that of insurance. I will insure my house against fire even though the premium is higher than the expected value of damage, because I am risk-averse on that scale and could not easily stand that much damage. The insurance company links me in a pool with thousands of other house owners, and makes a good living out of the difference between the sum of premiums and the sum of damages, which approximates to the statistical expected value of damage. A large company may consciously decide not to insure against some risks, where a single event would not bankrupt it, relying on the averaging or pooling effect to even things out over the years. In effect, the company does its own internal insurance and saves the difference that an outside insurer would have lived on. The idea that governments should behave risk-neutrally relies on similar pooling of risk, as if the government were a very large company. The trouble with this argument is that the risks and uncertainties of public projects are in reality not uniformly spread over the large number of people represented by the government. They are often likely to bear on just one section of the population, and then risk-aversion and non-linear preferences are relevant, just as for an individual. Another objection lies in the case of pure public goods such as clean air, where spreading the effect over a million people does not make its impact any less per person than if it only affected me. So it cannot be automatically assumed that CBA of public projects can use risk-neutral and linear methods.[7]

One way of dealing with risk-aversion or non-linear utility functions is to convert or map the usual CBA parameters onto a preference or utility scale, using a non-linear mapping or conversion or scaling function; in place of expected value one computes expected utility. This is done to some extent in decision analysis methods, such as multicriterion analysis (see Appendix F), but seldom in CBA. Another and more generally applicable way is to present probability information to decision-makers and let them apply their preferences, of whatever shape, directly (making sure they realise that the CBA applied risk-neutral methods so that any subsequent application of their own risk-aversion is not double-counting). If the non-linearity resides in the project's own characteristics, in the functions that connect input parameters, such as prices and quantities, to CBA indicators such as NPV, then risk analysis by simulation can help. This is discussed in Section A.3.

There are a number of formalised decision rules that may come close to matching decision-makers' preferences. They may help when competing projects or options have similar expected values but very different degrees of uncertainty. The *maxi-min rule* is very cautious and chooses the option that has the least bad worst-case, regardless of how much better some options might do in favourable circumstances; it uses only a small part of the available information. More subtle, but still on the risk-averse side, is the *minimax-regret rule*, which seeks the option that minimises the regret, where regret is the margin by which the chosen option is inferior to another option in any circumstances (crudely, the argument is 'I have less to lose by option X'). Both these rules refer primarily to decision situations where the uncertainty stems mainly from circumstances outside the projects or options themselves, such as demand growth or climate change. There are other rules for gamblers and some compromise rules that use all the information and allow for degrees of risk-aversion.[8] This whole approach can be called *scenario planning*, as it depends on defining a manageable number of scenarios (favourable, unfavourable, low demand growth, etc.) and then making pair-wise comparisons between outcomes for particular options under particular scenarios (e.g. by means of payoff matrices and regret matrices). It can help decision-makers by clarifying complex matters, but cannot remove the need to form or hold some opinion about the relative probabilities of different outcomes.[9]

One feature of any analysis of decision-making under conditions of uncertainty is that it highlights the value of *information*. Information usually comes at a price, and is measured not only in money or equipment or skilled people's effort but also in the passage of time. One may have to choose between making a decision now with very partial information and waiting two years for a pilot project or feasibility study, or just a rain gauge, to yield more comprehensive or more precise information. That choice itself must be guided by some sort of CBA, if only informally: How much better will the main decision be in two years' time, and is it worth waiting for? The choice can be guided by doing the CBA for the main decision with the information available now, and using sensitivity tests to show what extra information would be most useful to improve the decision. Such sensitivity tests can also help to redesign the project, either to make it better even in base-case situations, or to make it less prone to uncertainty (more robust or *flexible*). Sensitivity tests and common sense are a powerful combination for improving the design of projects. This use of uncertainty analysis to guide information-gathering activities is especially applicable in a prefeasibility study, which should, among other things, tell us what information to seek in the subsequent feasibility study.[10]

A.3 Ways of handling uncertainty in CBA

A.3.1 Crude adjustments to the best estimate

It is a common, but not a good, practice to respond to perceived uncertainty by biasing estimates in the direction of caution. Ways of doing this include:[11]

(a) *Risk premium on discount rate*
This practice, which is described in Appendix E (Section E.5), is borrowed from the financial behaviour of investors. It is only appropriate if the uncertainty is expected to increase with time,[12] and even then it is questionable. It applies a systematic bias against the longer-term effects of a project, whether costs or benefits. This thinking may be covertly included in the application of relatively high cut-off discount rates to certain countries.

(b) *Shortening the analysis period*
This is another approach which implies that events far in the future are particularly uncertain: it merely ignores them by terminating the analysis at a relatively small number of years. The method is crude, arbitrary, does not make use of all available information, and introduces a time bias that is nearly always inappropriate, except in a lower-bound estimate.

(c) *Biased parameter estimates; contingency allowances*
It is very easy to fall into the trap of 'adding a bit for luck' when making cost estimates with limited information. The extra bit, over and above the estimator's best-estimate or expected value, is often called a 'contingency allowance'. As stated in Section 2.2.5, this is sensible in a budget estimate, but inappropriate in a CBA because other parameters in the analysis, notably the benefits, are normally best-estimates. Nevertheless, many analysts do include such biased estimates in CBAs, and some authorities condone or even encourage the practice, generally as a crude antidote to a real or imagined tendency for estimators to be too optimistic. British government guidelines warn explicitly against 'optimism bias' and recommend explicit measures against it.[13]

None of these procedures is recommended. They obscure the true nature of uncertainty, thus reducing the desirable transparency of the analysis, and are liable to deceive or confuse the decision-makers.

A.3.2 Sensitivity tests

The principle and practicalities of sensitivity tests were explained in Chapter 3. They represent the most basic way of bringing uncertainty into CBA. In their simple form, when a parameter such as capital cost or annual benefit is varied by an arbitrary percentage, they say nothing about probability. The first way to make them yield more information to the decision-maker is to test not an arbitrary deviation from the best-estimate but a conceptually or physically meaningful one whose probability can be appreciated even if not statistically quantified (e.g. the effect of an earthquake similar to the worst on record, or an under-estimate of cost similar to a notorious one from the recent past).

If the analyst cannot fix on some specific deviation as being a meaningful one, it is usually best to quote switching values that enable the decision-maker to apply *lower-bound* reasoning to the interpretation of the analysis.

Interdependence or correlation between the variables or parameters that go to make up a CBA must not be forgotten. It has been recommended in Section 3.7 that, when sensitivity tests are done and reported, only one parameter should be varied at a time while the others remain at their base-case values. This does not imply that correlation is unimportant or absent, merely that a multiparameter sensitivity test is not a good way to examine it. An exception is the case when it is suspected that two parameters might be correlated and a joint test is done to find out to what extent they reinforce or coun-teract each other in determining the outcome of the analysis. But this needs to be very carefully reported, to avoid giving the false impression that an arbitrary test represents someone's estimate of a real correlation. Subject to such caution, it is generally good practice to report, alongside the numerical results of sensitivity tests, whatever information or indications are available about both probabilities and correlations.[14]

Occasionally, especially in prefeasibility studies or others at early stages of the planning process, the information base may be so sketchy that it is difficult to decide what parameter value within a plausible range should be named as the base case. Generally the analyst should name some value and explain his doubts in the report, but sometimes he knows that different decision-makers, or different sections of the report's diverse readership, will hold opposing views on the matter. In that case he may consider sitting on the fence and quoting several results alongside each other without naming a base case at all. This tactical retreat should only be used as a last resort.

A.3.3 Risk analysis by simulation
Risk analysis is the logical extension of sensitivity analysis when there is enough information to take probabilities and correlations into account

explicitly and numerically. It has developed largely in management and financial contexts, and its use in CBA has been discussed extensively since 1970, when Pouliquen and Reutlinger wrote a pair of World Bank staff occasional papers which showed that it was potentially useful, that analytical methods were largely impracticable, and that simulation methods were practicable, although time-consuming.[15] The fact that these two papers and their case studies are still being quoted 40 years later shows that the method has been more discussed than used, but the appearance since 1990 of cheaper and more powerful computers, and appropriate software, is changing that. The analysis needs more information than simple sensitivity tests, but is not difficult to apply. This section briefly describes the technique and its uses, and the reader who wishes to use it will normally find further guidance packaged with the relevant software.

Step 1 in applying so-called risk analysis to a CBA is to identify the key parameters whose variations have significant effects on the outcome: this can, of course, be done by sensitivity tests. In practice, it is not wise to apply the technique to too many parameters; six completely independent parameters is the suggested maximum, and a useful analysis can be done with about three (see below concerning correlated parameters). They might include a parameter related to climate change, or the depth to sound rock if that is not well known at the present state of study of the project, or the growth of demand for the project's main product or purpose (traffic for a road project, energy for a power station, etc.), or the general level of costs where this is hard to predict (as it is when the financial tender prices submitted by contractors depends largely on how desperate their market position is at the time). If escalation (gradual rise of real-terms price of inputs or outputs) is included in the analysis, it is usually wise to include it in the risk analysis, or at least examine it carefully by a sensitivity test to see if it is important. Each parameter used in the risk analysis should be a quantifiable one whose value is an arithmetic input into the calculation of the CBA indicator.

Step 2 is to estimate the probability distribution for each chosen parameter. Any method can be used, from sophisticated statistical analysis of past experience to educated guesses by experienced people. There are ways of drawing out (eliciting) such estimates, related to survey techniques and game theory. By putting carefully structured questions to the expert or decision-maker (or to an analyst who estimates the decision-maker's response) one can build up a reasonable estimate, remembering of course that high precision is impossible.[16]

155

Step 3, both a difficult and an important one, is to estimate cor-relations or covariance between the chosen variables. If there is any functional reason or causal link that might tend to make one parameter deviate from its expected value (in either direction) under the same circumstances that cause another parameter to deviate, the strength of that tendency must be estimated or guessed in terms of the usual correlation coefficient (+1 for total positive correlation (X up when Y is up), 0 for no correlation (X and Y independent), −1 for total negative correlation (X up when Y down); intermediate figures such as +0.7 and −0.4 for degrees of correlation). This applies either when one parameter is partially a cause of another or when they are both affected by a third phenomenon, a common cause, which is not one of the analysis param-eters. Examples of correlated parameters are crop yield and transport cost, or construction duration and construction cost, or two different sorts of cost. If the correlation is strong and systematic it may be prefer-able to bring the common cause into the spreadsheet and make it an independent variable. The total number of independent and partially dependent variables should probably not exceed about 10. Correlation can have a major impact on estimates of outcome probabilities, and should be included with care: either omitting or overstating it can invalidate the analysis.[17]

Step 4 is to simulate the analysis (run the model) a large number of times with different values of the chosen parameters each time. For instance, the complete CBA calculation is done 1000 times, deter-mining and recording the NPV (or other indicator) each time. The values taken by each varying parameter, over the whole series, must conform to its probability distribution as defined in Step 2, and if cor-relations are in force the cross-correlations between parameters must also be as specified. The amount of computation is immense, but modern personal computers can complete several hundred iterations in a matter of minutes, even for complex cases.

Finally, as Step 5, the results must be presented and interpreted. One or more output parameters, normally the CBA indicator such as NPV, will have been recorded for each iteration, and now the probability distribution of the values of the output parameter can be presented, as a histogram, as a cumulative curve or as a table of descriptive numbers, such as the mean, standard deviation, quartiles, deciles and extremes. Even when the varying input parameters have rectangular or very skew distributions, the output parameters tend to have relatively smooth and symmetrical ones, especially if the correlation is weak or not adequately incorporated. This is not a speciality of a particular CBA, but

156

a general statistical phenomenon for nearly all numerical processes. It is, however, quite likely that the expected value of the indicator will be significantly different from the value that would be derived from a single calculation using the expected values of all input parameters, especially if correlations are strong. Provided that the estimates are sensible this difference is not a delusion, and the risk analysis gives a better decision guide than the single result, even with sensitivity tests.

The last two steps (simulation and result presentation) are not practicable without a modern computer and appropriate software. The software can take the form of an add-on to a standard spreadsheet package. Once the CBA has been set up as a spreadsheet, which is the normal procedure anyway, it is relatively easy to install the add-on, replace single numbers with defined probability distributions for the few chosen parameters, define the correlations, and designate the output variables. The software allows any probability distribution to be entered, and offers a selection of standard low-information-content ones, such as the normal, log–normal and triangular distributions or truncated forms thereof. Any of the four distributions in Figure A1 could be entered very easily. The sampling methods whereby the computer ensures that the defined probability distributions and correlations are applied are complicated, but the user does not need to know about them. If a large number of values are taken at random the method is called *Monte-Carlo*; several hundred iterations are normally used, or even a thousand or two. To save time one can alternatively instruct the software to use a more subtle sampling method, such as *Latin Hypercube*, which selects values within each varying parameter's defined range in such a way as to achieve the desired distributions and correlations in fewer iterations than the Monte-Carlo technique would need. When it comes to the presentation of results, the software offers a wide choice of numerical and graphical formats suitable for inclusion in reports.[18]

Risk analysis of this sort is a powerful tool, and to use it casually would be like a weekend driver taking a racing car out for a spin, except that the first casualty is more likely to be truth or relevance than the user or a bystander. An analyst contemplating using risk analysis should experiment with the software and get a feel for what it does before applying it to an important analysis. The experimentation should include changing the correlation coefficients, the effect of which can be surprisingly strong. Although risk analysis is in other respects a step beyond sensitivity tests, it may be wise to do some sensitivity tests on the risk analysis coefficients. It is also useful to gain experience of the effects of having fewer or more parameters varying. If there are too many of them, and/or if the assumed correlations are too weak, their variations tend to cancel out in the computation and produce an unrealistically small spread in the output variable.

Although risk analysis was not used for CBA to any great extent before the mid-1990s, it is sometimes advocated as a regular procedure, notably in the World Bank, Asian Development Bank and UK government handbooks.[19]

A.3.4 Discontinuities, irreversibility and special risks

There is a danger that all this analysis of continuous variables and matters of degree will obscure some important absolutes and discontinuities. For instance, under some circumstances, which are not in the base case but have a significant probability of occurring, a project may have irreversible consequences. It might cause the irreversible loss of biological variety or a cultural site, irreplaceable soil loss following erosion, or radioactive contamination that would take centuries to decay.[20] There may be thresholds or other non-linear or discontinuous causal relationships that can have drastic consequences. These facts about a project or decision need to be either incorporated in the CBA or prominently reported alongside it. When a potential irreversible effect is significant, the '*option value*' of waiting for more information must be balanced against the cost of waiting.

One way of dealing with such effects is to use a veto or compulsion alongside a CBA, for instance in a multicriterion analysis as discussed in Appendix F. In environmental matters there are well-established concepts such as the *precautionary principle* ('assume the worst case scenario with respect to actions whose outcomes are uncertain ... other formulations abound), a *safe minimum standard* or a critical load, which in simple terms are centrally determined thresholds. All these can be applied rigidly or flexibly, with words such as 'reasonable' and 'acceptable'.[21]

Another approach sometimes discussed is *risk–benefit analysis*, which is a kind of CBA where the project being analysed is a way of countering a risk, such as a particular sort of accident. There is not actually any intention to implement the project, but the cost of its hypothetical implementation gives an indication of the economic value of the risk, that value being needed for the analysis of something else.

Acceptable risk analysis is a broad-ranging set of methods of arriving at safe levels of significant parameters, especially environmental ones. It combines several sources of information and opinion.[22]

A.4   Summary and recommendations

The treatment of uncertainty in and around CBA is a complex subject in which ideas are developing and changing, so that there is much to be gained by trying relatively new techniques. At the very least, every CBA should involve a few sensitivity tests, thoughtfully designed and carefully and transparently reported. The base case should incorporate best-estimates of all parameters, and any deliberate caution or pessimism in estimating should be identified and reported separately. Deviations from the base case should preferably be related to meaningful physical situations or thresholds rather than to arbitrary percentages.

For major analyses or projects with potential environmental or irreversible consequences, consideration should be given to carrying out a formal risk analysis using risk matrices and/or simulation methods. The analyst should, however, first gain experience in the use of the simulation tool, or draw on the experience of someone who has used it before.

All analysis and discussion of uncertainty demonstrates the value of information and the potential for feedback from analysis to design. An analysis of uncertainty done early in a project's development, even with crude estimates, may have great value in guiding the design or redesign of the project, perhaps incorporating conscious risk management features or a focused effort to achieve flexibility or robustness in critical areas. Investment in information, even at the expense of delay, may pay off handsomely.[23]

Flexibility is itself a valuable feature of a project: a project that is built in stages so that the design or even size of stage 2 can be adjusted at the last moment in the light of stage 1 experience, and of up-to-date demand forecasts, may be inherently better than a theoretically cheaper single-stage alternative.

A.5   Further reading and information sources

Most books on CBA have something to say about uncertainty, usually also using the term 'risk'. The notes to this appendix provide a number of references (full details of the works are given in Appendix K), although the reader should be prepared to find some confusion, especially regarding means, modes and medians. The most useful works for detail and discussion include:

- Schofield (1987) for risk-averse attitudes in a CBA context
- Watson and Buede (1987), for a wider discussion of decision-making under uncertainty

- Brent (1990) for risk analysis and the question of risk-aversion for public projects
- Goodwin and Wright (1991), for a clear discussion of probability functions, risk aversion, simulation, the value of information and the handling of uncertainty in a multicriterion context
- OECD (1995), for a balanced discussion from an environmental viewpoint, with emphasis on the value of information
- Abelson (1996, Section 2.7), for comments on expected values and risk aversion in the environmental context
- the British government handbook (H.M. Treasury 2003, in particular Chapter 4)
- the World Bank handbook (Belli *et al.*, 2001, Ch. 11), for a discussion of the limitations of sensitivity tests and the use of risk analysis
- papers of the Asian Development Bank, some of which are obtainable from the ADB website
- Zerbe (2008, Part IV), which reprints eight papers.

This being a field in which ideas and techniques are developing rapidly, the reader who wishes to find out more or to stay up to date should use a library or search the internet. Suggested keywords for index searches are: 'risk', 'uncertainty', 'probability', 'decision', 'game theory', 'Monte-Carlo' and 'precautionary principle'.

Notes

1 British government guidelines, however, give considerable emphasis to the consideration of uncertainty (H.M. Treasury, 2003, Ch. 5). The World Bank handbook covers risk and probability at some length (Belli *et al.*, 2001, Ch. 11). The Asian Development Bank (2002) issues separate documents on the subject.

2 More details, from the CBA point of view, are given in Irvin (1978, Section 3.03) and in Schofield (1987, Ch. 7). The importance of using means rather than modes for CBA is emphasised in Belli *et al.* (2001, Ch. 11).

3 Some people prefer to estimate damages at the boundaries of the probability slices rather than their typical floods as here. Another detailed example is given in Zerbe (2008, Part IV).

4 This was the formulation used in the 1991 version of the British government handbook, and an example is given in Box 13 of the 2003 version (H.M. Treasury, 2003). The outturns can even differ in direction (some positive, some negative); an example is given in OECD (1995, Section 9.2iii).

5 In the economist's terminology, the risk-averse curve shows diminishing marginal utility of money. For a fuller discussion see Schofield (1987,

Section 7.2.2) (an economics approach), or Watson and Buede (1987, Section 3.3.1) or Goodwin and Wright (1991, Ch. 4) (both a decision analysis approach), or Hertz and Thomas (1983, p. 185) (a risk analysis approach). Gamblers may show risk-seeking behaviour, represented by a concave-upwards line. People can be risk-averse at some levels and risk-seeking at others, represented by an S-shaped curve with concave and convex ranges. Good reasons for being risk-averse are given in OECD (1995, Section 9.3).

6 This derives largely from a seminal article in 1970 by K.J. Arrow and R.C. Lind (reprinted in Layard and Glaister, 1994, Ch. 4).

7 The 1991 British guide said '*some central government projects may impose large uncertainties on individuals – a cost which should be reflected in the appraisal*'. Distinguishing between the effects of a project on different groups of people within a nation is a major theme of the World Bank's approach: Brent (1990, Section 11.2.3) emphasises this aspect of the Squire and van der Tak approach, and Belli *et al.* (2001, pp. 163 and 207) explain why a government might undertake a project too risky for private capital. The World Bank handbook also examines gainers and losers at length (Belli *et al.*, 2001, Ch. 12). Potts (2002) also discusses impacts on particular groups.

8 The risk-seekers rule is called *maxi-max*, and two compromise rules are those of Hurwicz and Laplace. For further discussion see Schofield (1987, Section 7.3), Nutt (1989), Irvin (1978, Section 3.08) or OECD (1995, Section 9.4).

9 This is emphasised in the official British guide book (H.M. Treasury, 2003, Chs 5 and 6; a maxi-min example is given in Box 20).

10 OECD (1995, Section 9.1) calls this *investing in information*: it can be a very profitable investment by any standard. Delay in starting a project has option value; one of the best principles of decision-making is to keep your options open, and not to finalise a decision before you have to. The value of paying for more certainty is recognised in the official British guide to public sector appraisal (H.M. Treasury, 2003, Sections 33–38). Belli *et al.* (2001, pp. 163–164) reflect the World Bank's appreciation of the way uncertainty analysis can improve design.

11 These procedures are described and briefly criticised in Schofield (1987, Section 7.1).

12 Brent (1990, appendix to Ch. 11). Belli *et al.* (2001, p. 165) also considers a risk premium on NPV.

13 H.M. Treasury (2003, Annex 4).

14 A useful discussion of sensitivity tests is given in ODA (1988, Ch. 6). The World Bank handbook points out the danger of presenting one-parameter-at-a-time tests without mentioning correlations, and goes so far as to say that other techniques, such as risk analysis, are preferable to sensitivity tests (Belli *et al.*, 2001, Ch. 11).

15 Pouliquen's method and case study are summarised in Brent (1990, Ch. 11).

16 Watson and Buede (1987, Sections 3.3.2 and 7.4), Hertz and Thomas (1983, p. 30), Brent (1990; 1998, Section 11.2.2), Goodwin and Wright (1991), H.M. Treasury (2003, Chapter 4).

17 The importance of correlation/covariance, is emphasised in the more recent versions of the guidelines of development and funding bodies, such as Belli *et al.* (2001, Ch 11), Asian Development Bank (2002 and updates) and H.M. Treasury (2003, Annex 4). The smoothing effect of ignoring correlation is mentioned in Irvin (1978, Section 3.06).

18 Details and discussion can be found in the documentation of the software, such as that marketed under the name @RISK by Palisade Corporation (http://www.palisade.com/risk). This software, which is an add-on to widely available spreadsheet packages, can, of course, be used for other purposes besides CBA, such as cost estimating and almost any kind of forecasting. Other software packages, some available free of charge, can be found by searching the internet under 'Monte-Carlo analysis'.

19 Belli *et al.* (2001, Ch. 11), and also Asian Development Bank (2002 and updates), H.M. Treasury (2003, Annex 4). Refinements to Monte-Carlo analysis are discussed in Balcombe and Smith (1999). The use of risk analysis in a financial context, although without the benefit of spreadsheet add-on, is described in Yaffey (1992, Ch. 13). A wide discussion is given in Goodwin and Wright (1991, Ch. 6).

20 Irreversible effects are emphasised in the official British guide (H.M. Treasury, 2003, Ch. 4) and in Penman (1999, Section 3.2.3).

21 These are described in OECD (1995, Box 9.1). The difficulty of defining the precautionary principle is discussed in Abelson (1996, Section 2.7).

22 Both risk–benefit analysis and acceptable risk analysis are described in Winpenny (1991, Section 3.4.5).

23 It is no accident that these aspects are emphasised particularly by the more practical books on the subject, those issued by official agencies: see ODA (1988, Paragraphs 6.28–6.29 and 6.33–6.36), H.M. Treasury (2003, Ch. 4) and OECD (1995, Sections 9.1 and 9.5).

B

The domestic pricing and foreign exchange numéraires

B.1 Introduction

This appendix supplements the discussion in Section 2.3 of the two alternative definitions of the common unit of measurement for CBA, which is conventionally termed the *numéraire*.

An economic analysis normally seeks to measure a project's impact on the economy of a whole nation, for which purpose the financial prices obtaining in the market are not always appropriate; they have to be adjusted to bring them to economic prices. One reason for this, among several, is that foreign and local currencies are often not realistically related by the exchange rate officially applied.[1] There are two fundamentally different ways of making the necessary adjustments, based on two different definitions of the common unit of measurement. Section B.2 defines the relevant categories of goods, Sections B.3 and B.4 describe the two methods, Section B.5 summarises the differences, and Section B.6 discusses the choice between the two systems. For the more esoteric details, Section B.7 and the Notes guide the reader to some of the many books that treat this subject.

First, however, it is important to emphasise some aspects that are *not* involved here:

- These differences have nothing to do with time: the discounting procedure is the same whichever method is used, and the choice of discount rate is, for practical purposes, independent of the choice of numéraire.[2]
- The differences discussed here refer only to price distortions and corrections related to international trade and currency exchange rates: shadow pricing to deal with other distortions or to apply other criteria, such as the economic cost of unskilled labour relative to other domestic factors, is an entirely separate and independent matter whose factors and adjustments operate in addition to those discussed here.

B.2   Traded and non-traded goods

Goods and services that are traded across the border of the country for which an economic analysis is being done are called *traded goods*. These include imports and exports, and they may be relevant to a project in several ways, such as:

- project products directly exported, such as electric energy or manu-factured or agricultural products sold abroad
- project inputs, such as cement for a country that does not make enough of its own, or imported equipment and materials
- import substitutes, such as agricultural products which, although not physically exported, satisfy local demand that would otherwise have been satisfied by imports
- indirect traded components, such as the energy component of the cost of local manufacture of cement in an energy-scarce country that imports oil: there are practical limits to how far an analyst can go in tracing indirect traded components, but major ones usually need to be identified if the project has significant international trade impacts.

All the rest are *non-traded goods*, i.e. those that are generated and consumed within the country. They include both goods that have market prices and those that do not, such as public health improve-ments, reductions in road accidents or education. The methods for putting a price on the latter are the subject of Appendix C; in this appendix it is assumed that every non-traded good has either a local market price or an equivalent price arrived at by some estimating method.

Sometimes it is useful to subdivide the non-traded goods further into those that, by their nature, cannot be traded internationally (*non-tradeable goods*) and those that could be traded but currently are not.[3]

B.3   The domestic pricing system

The simplest economic numéraire, and the one usually appropriate to projects whose inputs and outputs are mostly produced and enjoyed within one country, is the *domestic pricing numéraire*. This is sometimes called by other names, such as the *willingness-to-pay numéraire* or the UNIDO system.[4] The precise definition is that all costs and benefits are valued in terms of their impacts on the society's consumption, measured as what people are willing to pay for each good or service (at the margin, i.e. the willingness-to-pay for one more unit).[5] For practical purposes this means that the numéraire is market prices inside the country, although with adjustments for market distortions.

For a non-traded good, the valuation is straightforward: its financial market price is also its economic price with this numéraire, unless it needs to be adjusted for some other reason not connected with international trade and exchange rates. If a cubic metre of locally produced timber costs Rs 12 000 on the local market, that is also its economic value.

For a traded good, however, the financial market price is not a good measure of economic value under this numéraire, because it is at least partly conditioned by the unrealistic exchange rate. Suppose a project's construction uses imported cement for which the nation pays US$150 per tonne at the border, and the *official exchange rate* (OER) is Rs 30 per US dollar: the financial cost of cement will be Rs 4500/t (plus transport and handling costs, but they are non-traded items and will be omitted from this discussion). Suppose that the analyst estimates the effective or realistic exchange rate to be Rs 40 per US dollar. Then the estimate of the economic price of cement in terms of this domestic pricing numéraire is the border price multiplied by this *shadow exchange rate* (SER); thus $150 × Rs 40/$ = Rs 6000 per tonne.[6]

B.4   The foreign exchange system

The other economic numéraire is the *foreign exchange numéraire*, also called the OECD or LM or ST or LMST numéraire, or the world price system or level.[7] The numéraire is strictly defined as free foreign exchange in the hands of the government of the nation for whom the analysis is being done. The unit of calculation can, in principle, be a convertible foreign currency such as US dollars, but it is usual for the values of costs and benefits to be expressed in local currency using an official exchange rate, e.g. in foreign-exchange-equivalent rupees (sometimes called *border rupees*).

Under this system the valuation of traded goods and services is straightforward: the tonne of imported cement would be valued at the border price converted into local currency at the official exchange rate (OER): $150 × Rs 30/$ = Rs 4500 per tonne.

The correction for the unrealistic exchange rate is this time applied to the non-traded goods: in the simple form of this method their local market prices are multiplied by a *standard conversion factor* (SCF), which is the ratio of the official and shadow exchange rates (with an overvalued local currency this ratio is less than one). The SCF is the ratio by which the local currency is judged to be overvalued, OER/SER; in the above example it would be 30/40 or 0.75. The cubic metre of locally produced timber, costing Rs 12 000 on the local market, would thus be valued at Rs 9000 in the economic analysis. In a more subtle form of the foreign exchange analysis there would be different conversion factors for particular goods or groups of goods.[8]

B.5   Comparison of the two systems

In a country with a freely convertible currency, foreign exchange can at any time be converted to and from the money held by the citizens; the official exchange rate is a realistic one, and there is then little practical difference between the two numéraires. In a country with significant foreign exchange or trade restrictions, however, this is often not the case, and the two numéraires cause goods to be given numerically different values, even if expressed in the same currency. The above examples of the valuation of imported cement and local timber illustrates this. If both methods are used in their simple forms, without distinguishing groups of goods with specific factors or adjustments, the economic values are as in Table B1.

It is clear that (in this simple case with a single SPF) the domestic pricing system gives rise to larger numbers than the foreign exchange system by a consistent ratio of 1/0.75, which is the ratio of the exchange rates (the domestic pricing numéraire is a smaller unit so more units are needed to express the same economic value). As the same ratio affects all costs and benefits, the effect on economic indicators will then be:

- total benefits, B, total costs, C, and their difference, NPV, will all be bigger numbers in the domestic pricing system than in the foreign exchange system, but a positive NPV in one system will always correspond to a positive NPV in the other
- any ratios between economic values, including any B/C ratio, will be the same in both systems
- IRRs, as they are defined by the transition between positive and negative NPV (or by B/C = 1), will be the same in both systems.

If the analysis is a more subtle one with disaggregation of particular types of goods and services and the application of different adjustments to each group, such as group-specific conversion factors instead of the SCF in the foreign exchange system, then the above statements about equivalence of indicators will not apply precisely.[9] The indicator values may, however, still be similar, and the outcome of comparisons

Table B1 Comparison between the domestic pricing and foreign exchange systems

System of valuation	Domestic pricing system	Foreign exchange system
Other names sometimes used	UNIDO, willingness-to-pay, domestic price level	OECD, LMST, SCF-system, conversion factor approach, world price level
Valuation of *traded* goods (one tonne of imported cement)	Convert *border price* at shadow rate SER ($150 × 40 Rs/US$ = Rs 6000)	Convert *border price* at official rate OER (US$150 × 30 Rs/$ = Rs 4500)
Valuation of *non-traded* goods (one cubic metre of local timber)	Use local market price (Rs 12 000)	Convert local market price by SCF = OER/SER (or specific CFs for particular goods) (Rs 12 000 × 0.75 = Rs 9000)

between projects, or between one project and predefined test or cut-off values of the indicators, may still be the same. Thus the choice of numéraire is seldom likely to change the final recommended decision that emerges from a CBA. People use one system or the other from habit, to achieve consistency with some other analysis, or because of a perception that trade distortions are important.

It is important to remember that border pricing and world prices are used for traded goods in both methods, being the correct measure of national opportunity costs. The names *world price level* and *domestic price level* must not lead people to believe otherwise.

B.6   Choice of system

In most countries with convertible currencies the domestic pricing system with its willingness-to-pay numéraire is normally used, and the distinction is often not mentioned at all in reporting an analysis.[10] In countries where international trade is restricted and distorted, foreign exchange is scarce, the local currency is overvalued by the official exchange rate, and/or there are significant differences between the official rate and one or more free or black market rates, the foreign exchange numéraire is sometimes preferred. The tendency for the less developed countries to have unrealistic exchange rates and overvalued local currencies has become much weaker since the 1960s and 1970s, when the foreign exchange approach was developed, so the case for this numéraire is less compelling than it was.

In essence, the foreign exchange approach takes an outsider's view and sees the local currency as overvalued and needing adjustment, while the domestic pricing approach takes an insider's view and sees foreign currency as undervalued by the official exchange rate, so this is corrected by using the more realistic shadow exchange rate. The domestic pricing numéraire is easier for local decision-makers to identify with, because most of the prices are those encountered in the market; most people find it easier to grasp that the prices of imported goods are adjusted upwards to allow for scarce or undervalued foreign exchange than to grasp that local prices are adjusted downwards to allow for an overpriced local currency. In the domestic pricing system, the economic prices of those goods that do not need adjustment to correct particular distortions are prices numerically the same as their financial prices, which is intuitively sensible and means the economic analysis is done in a unit that means something to local report readers; the upward shadow pricing of foreign exchange sits comfortably alongside the upward adjustment of other scarce goods and the downward adjustment of overplentiful goods such as unskilled labour. In the foreign exchange system expressed in local currency, by contrast, the economic analysis is done in a unit which is, by definition, distorted, because it depends on an unrealistic exchange rate. The unit is correct and consistent within the analysis for which it is defined, but is not easily related to other contexts. So the foreign exchange system is less transparent.[11] The one case in which the foreign exchange system has a clear advantage is where there are significantly different trade distortions for particular goods or groups of goods, and when different conversion factors are therefore used rather than a uniform SCF. Because of the exact correspondence of the numbers when standard factors are used for all non-traded goods, there seems little point in using the foreign exchange system unless such group-specific conversion factors are going to be used, and even then the same effect can be achieved easily with the domestic price system.[12]

International funding agencies such as the World Bank and the Asian Development Bank use both numéraires on different occasions. Until the mid-1990s they tended to favour the foreign exchange approach, but the 1996 draft handbooks and guidelines of both those institutions recommended the domestic pricing system, for reasons similar to those given here; and later revisions such as Belli *et al.* (2001) do likewise. Border prices must, of course, be used for traded goods, with either numéraire.

The recommended procedure for readers of this book who find themselves in the position of analyst is:

- ask the decision-makers, or the client who commissioned the CBA, which numéraire is wanted or preferred
- unless this produces a strong case for the foreign exchange (LMST) numéraire, use the domestic pricing one.

The discussions in this book, and most of the examples, use the domestic pricing system, both for the above reasons and because it makes shadow pricing easier to explain. It is important when conducting any CBA in an international setting to state clearly which system or numéraire is being used.

B.7   Further reading and information sources

This matter of the two numéraires has given rise to some confusion, some controversy, and much writing on the part of economists since the late 1960s, and indeed the fact that the two were in general mutually convertible was not initially realised by many users and commentators.

The following books, listed in chronological order, contain useful discussions of the subject, although many other books also deal with it (full references are given in Appendix K).

- Squire and van der Tak (1975, especially pp. 28, 35, 57, 93, 135), this World Bank publication first brought the two numéraires into a systematic comparison, and also developed the foreign exchange system for practical application.
- Irvin (1978, especially pp. 63–67, 75–77, 86–91).
- Gittinger (1982, especially p. 244), this book is widely used as a guideline, not only in its primary context of agricultural projects.
- Ray (1984), strong on the foreign exchange system.
- Brent (1990, 1998), a university-level economist's textbook devoted mainly to the foreign exchange system but including a thorough comparison with the domestic pricing system and some variants.
- Ward and Deren (1991, especially Sections 4–7, 10), this is a World Bank publication.
- Curry and Weiss (1993; 2000, especially Chs 5, 6), a comparison, with an example, and pricing of particular types of goods is discussed for each system.
- van Pelt (1994, Sections 3.4 and 3.5), includes an example worked both ways.
- Perkins (1994, Ch. 9), explains both systems and the situations when one or the other has advantages.

- World Bank (1996) and Belli *et al.* (2001), unlike most earlier publications by or linked with the World Bank, these unequivocally favour the domestic pricing approach (they use the term *conversion factor* for what the present book calls SPF, as distinct from the foreign exchange system's factor of the same name; see Chapter 5 of either of those books).

Notes

1 This may be due partly to explicit foreign exchange controls or an officially fixed exchange rate, but it may also be caused by trade restrictions or by taxes or subsidies on imports and/or exports; in general it is the consequence of instruments of trade policy. See Belli *et al.* (2001, Ch. 5).

2 The books that originally presented the two methods included slightly different approaches to the choice of discount rate, and some contemporary writers still maintain the distinction.

3 See Belli *et al.* (2001, Ch. 5); this World Bank handbook consistently uses the terms 'tradeable' and 'non-tradeable' rather than 'traded' and 'non-traded'.

4 The method's best-known exposition is in the UNIDO manual of 1972.

5 See Ward and Deren (1991, Ch. 4), van Pelt (1994, Section 3.5) or Perkins (1994, Ch. 9).

6 For a thorough analysis the estimate would be based on the world and local prices of the main traded items, using weighted averages. For details of the rigorous estimation of the SER see Ward and Deren (1991, pp. 36 and 41–43), Curry and Weiss (1993, 2000), Irvin (1978, p. 85) or Belli *et al.* (2001, Ch. 5). Even if foreign exchange dealing is unrestricted, any taxes, duties, subsidies or quotas applied to imports or exports will make the shadow exchange rate diverge from the market exchange rate. Where exchange dealings are restricted, a very rough guide to the realistic exchange rate (or at least to the direction and degree of adjustment needed) is the informal market price of foreign exchange: if informal money changing is illegal, this black market rate may itself be distorted by the element of risk to the participants, the level of risk being very different in different countries.

7 The system was first propounded in 1968 in a manual written for the OECD by I.M.D. Little and J.A. Mirrlees. Little and Mirrlees presented it again in a 1974 book in their own names, and there followed a period of debate about the relative merits of their procedure and the UNIDO one of the 1972 UNIDO manual. In 1975 the Little–Mirrlees (LM) methodology was modified and presented by two World Bank authors, L. Squire and H.G. van der Tak (ST, or S and T); their synthesis is often called the LMST system as an alternative to referring back to its origins in the OECD manual. See Squire and van der Tak (1975), Irvin (1978), Ward and Deren (1991, Section 5), Curry and Weiss (1993; 2000, Ch. 5) and van

Pelt (1994, Section 3.4). Teaching texts on the LMST method are given in Brent (1990) and Perkins (1994).

8 The term *conversion factor* has usually been used in connection with the foreign exchange numéraire, as here, but is sometimes (notably in the World Bank handbook (Belli *et al.*, 2001)) used to denote what this book calls a shadow price factor (SPF).

9 This use of several different conversion factors was the intention of the original proponents of the foreign exchange system, and is still advocated by some people, but it is seldom done in real analyses.

10 For the UK, the Green Book (H.M. Treasury, 2003) requires the domestic pricing system but does not explicitly name it as willingness-to-pay or UNIDO.

11 Most of these arguments in favour of the domestic pricing system are in an unpublished 1993 paper by C.E. Finney, and similar arguments are used in the World Bank handbook (Belli *et al.*, 2001, pp. 40–41). The circumstances when each system may be appropriate are discussed in Ward and Deren (1991, Section 7) and Perkins (1994, Ch. 9).

12 Ray (1984) emphasises the need, when using the foreign exchange system (which he calls the conversion factor approach), to use it thoroughly.

C

Ways of estimating economic prices

C.1 Introduction and underlying principles

This appendix supplements Chapter 2. Its purpose is to explain briefly how economic prices are determined, to introduce some of the technical terms that readers may meet in other texts, and to indicate where the reader can find further information. When economic prices differ numerically from financial or market prices, for whatever reason, they are often called 'shadow prices', or sometimes 'accounting prices'.

In the economic theory underlying CBA, the relative values to be put on different things are intended to reflect the aggregated preferences of individuals.[1] For the aggregation of costs and benefits, it has to be accepted that the preferences of different individuals can be added, and that the current distribution of wealth is acceptable. (This is because ability-to-pay obviously affects willingness-to-pay, although reservations about this latter assumption can be crudely taken account by income weighting, as described in Appendix D.) These two value judgements are debatable, but for the purposes of this appendix they are accepted.

For people's preferences to be used as a basis or an indicator of value, they have to be revealed in some way. People reveal their preferences in many ways, one of which is by how they spend whatever money they have. Another way is how they vote in elections, and yet another is how they answer questionnaires. This appendix describes briefly a number of analytical methods for turning some of these varied and often problematic expressions of people's preferences into numerical economic prices. All the methods are approximate, debatable and open to challenge. To choose and use any one of them is a value-laden decision, but so would be a decision to omit some class of costs or benefits because it was difficult to put a value on it.

Cost–benefit analysis – A practical guide
ISBN: 978-0-7277-4134-9

A relatively simple case is a good for which there is a market; the market price is a revelation of people's preference for that good relative to others, although that price may need adjustment. This is the subject of Section C.2. Another principle underlying economic pricing is that of opportunity cost from the point of view of a nation; some methods based on this approach are described in Section C.3. Section C.4 describes ways of using stated preferences derived from surveys (contingent valuation and contingent ranking). Section C.5 covers a number of other approaches and techniques relying on preferences that people reveal by their behaviour in matters related more or less indirectly to the good in question. Section C.6 describes how shadow prices can be used as instruments of policy, by governments which claim (with or without justification) to be applying inherent values or the preferences of their electors or subjects. The question of which method to use, when one has any choice, is touched on in Section C.7, and Section C.8 covers the transfer of estimates from one context to another. Finally, Section C.9, supplemented by notes throughout the appendix, points the reader to sources of further information and discussion.[2] To clarify the terminology used in this appendix and the literature it cites, some relevant definitions are summarised in Box C1.

In the case of environmental impacts or goods, it is often helpful to begin by linking the impact to some other good that is easier to value. The technical means to this end are called *dose–response functions*, or *damage and response functions*. These depend on finding and quantifying a relationship between the environmental or other effect being valued and some marketed commodity, such as agricultural production, or some effect that can be valued by one of the non-market methods set out below. Linkages to marketable goods are especially suited to valuing pollution which affects crops or buildings. The relationship to real markets makes such valuations appear relatively objective and precise, but in practice it is difficult to quantify the relationship or function with any accuracy, mainly because damage and production are affected by many factors at the same time. Another common application is pollution affecting health: for instance, medical knowledge can quantify the link between particulate matter from diesel exhaust and respiratory disease, and hence morbidity and mortality (the technical link), and these can then be valued by human capital or other methods (the valuation stage). The analysis needs considerable amounts of data and often involves regression analysis, linear programming or quadratic programming, but in some fields the data are available and the technical linkages may be already well established.

Box C1 *Types of goods and values*

A *public good* is one that cannot be withheld from one individual without withholding it from all (*non-excludability*), and consumption by one individual does not reduce the amount available for consumption by others (*non-rival consumption*). As well as pure public goods, there are many that approximate to this type without fitting it precisely. Private businesses are unlikely to provide public goods because they cannot arrange for users to pay and non-payers to be prevented from enjoying the service; public goods tend, therefore, to be either pre-existing conditions or goods provided by the state and financed out of taxes. Examples are good views, clean air, clean rivers, beaches or streets, street lighting, police protection, lighthouse beams, and the existence of whales, although I have never seen one.

A *private good* is one whose enjoyment can be restricted to people who have paid for it so it can be provided by a private sector business and marketed. Examples are potatoes, beer, concerts, paintings, zoos, metered water supplies and toll roads.

Many goods do not fall neatly into either of these two categories, and ways can often be found to make public goods into private goods, at least to some extent. Examples are private beaches or parks (with restricted access), amenities in private housing estates, healthcare and land drainage. Intermediate cases are sometimes called 'quasi-public' and 'quasi-private' goods. A more subtle classification recognises, between public goods and private goods, *toll goods*, which show a degree of non-rival consumption (or low subtractability) but not non-excludability, and *common property goods*, which show non-excludability but not non-rival consumption. Examples are telephones and irrigation schemes, respectively.

A *merit good* is one that society collectively decides to interfere with the provision of, rather than leaving it to individual choice. Examples are the arts, provision of minimum housing standards and health services. They can be public goods or private goods. Demerit goods are those which society disapproves of, although individuals choose them (e.g. substance abuse).

Use values are associated with benefits gained from actual use. Subsets include:

- *direct use values* – can be consumed directly (e.g. jam, petrol, books)
- *indirect use values* – functional benefits (e.g. flood control).

Non-use values (or passive use values) concern goods that an individual is not about to enjoy directly but still places a value on. They include:

- *option values* – someone values the option to see a whale or visit the Grand Canyon one day, without having a definite plan to do so
- *bequest values* – for keeping options open for future generations
- *existence values* – an altruistic desire that something, typically an environmental asset, should continue to exist.

Values of different kinds can often be added: for environmental goods it is generally true that: total value = actual use value + option value + existence value.

These classifications are not hard and fast (some people regard option values as use values, for instance), and overlaps or debatable boundaries are always possible, but they can be useful to clarify valuation methods. Environmental or finite-resource effects may mean that an apparent private good, such as petrol, is not a pure one. Goods are sometimes classified into use and non-use goods, as with values here. (Option and existence values are discussed in detail in Pearce and Turner (1990, Ch. 9). Public goods are defined and discussed in Perkins (1994, Ch. 12), Turner *et al.* (Coker and Richards 1992, Ch. 5), Penman (1999), Belli *et al.* (2001, Appendix 1A) and Brent (2006, Ch. 6).

C.2   Market prices with adjustments

In simple cases the aggregated individual preferences are revealed as willingness-to-pay for goods and services in a market situation. The market price is not, however, the sole component of willingness-to-pay, because there is also *consumer surplus*. When I go to buy fuel for my car I may be willing to pay 80 p/litre, in the sense that I would decide to do without the fuel if I could not find any priced at less than 81 p. But if competition between filling stations is at least partially effective and I have time to shop around, I may well find fuel at 71 p/litre. In that case my willingness-to-pay is still 80 p but the market price is 71 p; the 9 p difference is called 'consumer surplus'. The measure of my welfare from the litre of fuel is 80 p, of which I pay 71 p and enjoy 9 p as consumer surplus. Some motorists might be willing to pay more than me, and some less, but anyone not willing to pay the market price is no longer motoring, so the consumer surplus varies between consumers, from zero for the motorist who would rather stop motoring than pay 72 p/litre, to the other extreme of the one who would continue even if fuel cost several hundred pence a litre. The measure of total welfare enjoyed by all the motorists together is the summation of each one's willingness-to-pay times the amount he uses; in aggregate this may be much more than the amount paid for fuel.[3]

Provided that a particular project has only marginal effects on the market as a whole, the extra units of a good produced by the project will produce only a small relative change in the market price. Under those conditions the extra bit of consumer surplus will be negligible, and the market price multiplied by the extra quantity sold will be a good approximation to the extra welfare. (This applies to goods used by the project as well as to goods it produces.) So, in these circumstances, a free market price is an approximation to the aggregate marginal willingness-to-pay, the error due to ignoring consumer surplus being negligibly small.[4] Welfare changes can then be measured, to an acceptable degree of approximation, by quantity multiplied by market price. This is the basis for using money as a common unit of measurement for CBA.

Given that the numéraire is domestic prices, the market price of a good or service is the measure of the marginal willingness-to-pay, and therefore of its unit economic value, provided that the market approximates to a free market, with perfect competition and undistorted operation of supply-and-demand mechanisms. This proviso means that there are many independent producers and many consumers for the good in question (so no one can exercise monopoly power or single-buyer power), and that all participants are adequately informed. Another condition is that the good should be a private good such as petrol or jam, rather than a public good such as clean air. When these conditions hold, to a reasonable degree, the financial or market price is an acceptable measure of the economic price (i.e. of the unit welfare in terms of the chosen numéraire). The economic price is not the same thing as the financial price, but it is numerically equal to it, for this class of goods only; in terms of the arithmetical procedure given in Section 2.4, the shadow price factor (SPF) is 1.0. The economic price still measures value to society, not money.

Economists often refer to any departure from the ideal market conditions as 'market failure'. Market failure may be caused by monopoly (one-seller) or one-buyer situations, collusion between producers or between consumers, imperfect information, or regulations preventing

176

free-market conditions. Governments can compensate for market distortions to simulate a perfect market, but they can also cause market failure themselves, and they often do. Some economists call inappropriate corrections or distorting interventions 'government failure', and the extent to which they regard it as a bad thing is very variable.[5] Typical government distortions are protective tariffs, import and export bans or quotas, and selective taxes and subsidies.

It was explained in Section 2.2 that taxes, duties and subsidies are not economic costs because they are transfer payments. Conceptually it is possible to separate all transfer payments into one category in the financial analysis, and then omit that category in the economic analysis (or give it a shadow price factor of zero). This method is used in the explanation of shadow pricing in Section 2.4. It must, however, take account of the transfer payment elements of *all* cost categories, and most costs will have at least some tax or duty component: wage costs are partly tax, and many materials have an import duty or value-added-tax component in their financial prices. Instead of bringing all the transfer payments together in one category in the financial analysis, as in Section 2.4, it may be easier to achieve the same arithmetical effect by estimating the proportion of the financial price of each cost category that represents transfer payments, and then allow for those payments to be omitted from the economic analysis by a suitable SPF. For example, if it is estimated that 23% of the financial price of imported cement goes to taxes and duties, and if there is no other reason to apply an adjustment, the analyst may apply an SPF of 0.77 to all cement costs, merely as an arithmetically convenient means of removing the transfer-payment element. If, in addition, there were thought to be market distortions requiring an adjusting factor of 1.1, a combined SPF of 0.85 (0.77 × 1.1) could be used. Going a step further, it might be convenient to compute an overall SPF for reinforced concrete, which would be a weighted mean of the SPFs of the elements of the financial concrete price, such as cement, steel, labour, transport, etc. (These SPFs are sometimes called *conversion factors*, but that terminology is avoided in this book to avoid confusion with the different kind of conversion factor used with the foreign exchange numéraire; see Appendix B.)

Going beyond the removal of transfer payments, the derivation of shadow prices or SPFs to adjust for market distortions is a job for an economist and beyond the scope of this book.[6] In most countries, with reasonable approximations to a free market, such adjustments are not considered necessary at all.

C.3 Opportunity cost approaches

For some goods and services the opportunity cost approach can be used fairly directly to arrive at a reasonable estimate of the economic unit value or price (opportunity cost is explained in Section 2.5). This section gives some examples.

Labour

In developed countries the labour market is sometimes fairly unrestricted, and shadow pricing of the labour component of costs is seldom considered necessary. In developing countries, however, the market price for labour, especially unskilled labour, is sometimes a poor guide to the economic price because of minimum-wage legislation, rigidity in the labour market, restrictive practices, or the use of unpaid family labour in small enterprises and farms. A better guide to the economic price can then be obtained from the opportunity cost principle, but it must be applied carefully and realistically. The economic cost to the nation of employing an unskilled labourer on a rural road project is the marginal production foregone (or, in more general terms, the welfare foregone) that she would have produced if the project had not been there. In an agricultural setting that opportunity cost obviously varies very markedly with the seasons: at the peak of the harvest season the nation would lose a great deal by taking a labourer away from the harvest (part of the crop would be lost), but in the off-peak season, when his alternative occupation was non-urgent house maintenance, resting or a leisure pursuit, the cost to the nation would be very small (but not zero, as he is part of the nation and both his house and his leisure have value). The opportunity cost of employing him in an urban project is more difficult to estimate; migrant labour involves incidental costs, and sometimes two or more people may migrate to a city for every one job that is created, but still the marginal product in the rural situation from which labourers came is the starting point.

So the economic price of unskilled labour, when the effective alternative occupation is agriculture, is measured by the lost agricultural production and is likely to vary from one season to another. In some situations the going wage rate in rural areas reflects this, varying seasonally, so the financial price already includes the seasonal effect. Often, however, labourers are employed steadily through most or all of the year, and there is only one financial wage rate. As a rough guide, the economic price is likely to be near the financial rate in peak times such as harvest (SPF = 1, perhaps a little higher in acute peaks), and significantly lower (SPF = 0.4 to 0.8) in slack times. If unskilled labour represents a major part of a project's costs, or is crucial to a least-cost or choice-of-technology decision, it may be desirable to use different economic prices at different times of year. (This refinement is particularly needed in agricultural contexts and in less-developed countries or regions, and is accordingly dealt with in some detail in appropriate books or guidelines, such as the Asian Development Bank's guidelines, and also in Potts (2002) and in Curry and Weiss (2000).)

When there is a legal minimum wage and significant unemployment, the economic price may be approximated by multiplying that artificial rate by a low SPF, but it can also be guided by an unofficial (possibly illegal) going wage, which may represent something like a free market. The shadow or economic price of labour should be applied to all labour used in a project, including unpaid family labour if it is one of the inputs contributing to the project's production. An example is given in Appendix I.

For any skilled or scarce type of labour, however, the market price (the going rate) is usually a fairly good guide to the economic price, and in practice the SPF is usually taken to be 1.0, although one can make a case for a higher value for crucial skills or at times of peak demand.[7]

Land

A project often uses up land, which then of course needs to be valued for an economic analysis. If there is a relatively free market in land of the relevant type, the market price may be a sufficiently good estimate of the economic price. If there are evident distortions, an economist may be able to quantify them and apply corresponding corrections or adjustments, in effect simulating a free market. If there is a market for rented land, the expected rental value can be discounted to provide an equivalent lump-sum cost at the start of the project, or alternatively the rental value of the land can be entered into the CBA calculations as an annual cost for the project's whole lifetime.[8] If there is no market in the relevant sort of land, for instance if it has belonged to the government for a long time, the best guide to the land's economic value is a consideration of what it would or might otherwise be used for; by the opportunity cost principle this is the economic value. If it is agricultural land, or potentially could be, the annual net return (value of agricultural produce minus production costs, as in Section 4.2) can be estimated in economic prices to produce an imputed economic rent, and this can either be applied as an annual cost or be discounted to produce a one-off unit economic cost of land. This procedure should only be used when there is no market price, as market prices usually take good account of all kinds of value, such as environmental and non-use values, as well as the use value covered by agricultural production. At the least, when no directly relevant market exists, estimates derived from production estimates should be compared with market values somewhere else.

In the case of a project where land is a major input, such as an agricultural project, the land used may be dealt with by defining the without-project situation so that it covers the same area as the with-project situation. Then no explicit valuation of land is needed, as its opportunity cost is effectively subtracted when the incremental benefits are calculated from the difference between the two situations.

Different parts of a project's land-take may need to be treated separately, and in some analyses the valuation of land becomes quite complicated, although still resting on the same simple principles. If people are moved from land that is to be used for a project, resettlement costs must be included in the project analysis. Financial compensation costs for land must not be included as well as an economic opportunity cost for the same land, which would be double-counting. However, if the financial compensation includes an element to cover distress or cultural loss, that element should be included somehow; the excess of financial compensation over economic opportunity cost may give an indication of the size of that element, by the willingness-to-accept principle. An analyst can often obtain specific guidance on such details from the manual or guidelines of the appropriate decision-maker or loan agency.[9]

Imports and exports

Goods that are, or could be, traded internationally (*tradeables*) need special care. As explained in Appendix B, their opportunity cost from the nation's point of view is the world or border price converted into the domestic price numéraire by an effective or shadow exchange rate (which may differ from the official exchange rate), and adjusted for transport and handling costs. These adjustments must be made to or from the place where the goods produced or used by the project would be substituted for the equivalent goods from another world source. For instance, a project at Awash in Ethiopia was expected to produce cotton lint for export, and maize as a substitute for foreign maize that would otherwise have to be imported; it would also use imported fertiliser. For the economic analysis the price of the export at the project site was arrived at by taking the expected world price (which for cotton is quoted at northern European ports) and subtracting the costs of getting the project's cotton lint from Awash to a northern European port. The imported fertiliser was valued by taking the forecast world price at a north-western European port and adding the transport and handling costs of taking the fertiliser from such a port to Awash. (The prices of commodities are forecast some years into the future by bodies such as the World Bank, and the figures are quoted at some defined place appropriate to world trade patterns in the individual commodity.) For the maize, which would be a project output, but an import substitute, the transport and handling costs were added, as with the real import (fertiliser), not subtracted as for the other output (cotton). If the hypothetical place of substitution between the project's produce and world produce which it would replace or displace is neither the place where the forecast is quoted nor the project site, but a third place such as a factory elsewhere in the project country, some transport costs will have to be added and others subtracted, as exemplified in Appendix I.

A similar logic guides the application of quality premiums if the goods are of higher or lower quality than the grade used in the world price forecast. The prices at the project site derived using such calculations are called *import and export parity prices*.[10] The economic analysis of a project may require computation of values for several tradeables; the Awash project involved not just the three goods mentioned above but also a total of two exports, four import substitutes and two imported inputs.

Fuel wood

Fuel wood is mentioned here as an illustrative example to show how the consequences of something's production or use have to be thought through. In analysing an irrigation project at an underpopulated site in Africa, it was proposed that some of the better soils be used to grow tobacco, while the poorer soils would be used to grow fuel wood. Some of this fuel wood would fire the furnaces in the tobacco curing barns, and some would be burned by labourers and their families for cooking. In the economic analysis no specific benefits or costs related to the area under fuel wood were counted. The feasibility report stated that the cost of growing, cutting, transporting and handling the wood for the barns was included in the economic crop margin calculations for the tobacco, while the labour needed to bring fuel wood to the homes was compensated by the benefit enjoyed by the consumers, which in turn would help to attract labourers to the area.

That was a case where the production and use of the fuel wood was a local affair without implications for international trade (except through the tobacco crop). In some cases fuel wood can have trade implications by reducing the need to import oil for cooking, and if it replaces animal dung formerly burnt for cooking it may also reduce fertiliser imports (the dung being diverted for use as a fertiliser). So the valuation of the fuel wood would include calculation of the border effects of the changed importation of two other goods (oil and fertiliser) which it can substitute for, either directly or by way of an intermediate good (dung).[11]

Growing or using fuel wood can also have significant environmental effects, which may need to be valued as well, using the techniques described in Sections C.4 and C.5. As always, the key to correct and defensible valuation is the clear definition of the without-project situation (see Section 2.1).

C.4   Stated preferences; contingent valuation

When a good has no market price that can be used or adjusted, the economic price has to be estimated some other way. It is mostly environmental aspects and impacts, positive or negative, that have to be valued without reference to market prices, because no markets exist for the goods themselves. The most versatile valuing method, although one with many problems, consists of making a survey of a number of people, using a carefully designed questionnaire, and deducing the economic prices and values from the responses. The main class of methods is called *contingent valuation*, with *contingent ranking* as a subclass.[12]

Most environmental goods are public goods and have no market; people do not pay for enjoying them, or not directly. In terms of the breakdown of total willingness-to-pay (WTP) into actual payment and consumer surplus (see Section C.2), there is little or no payment and all or nearly all the WTP is made up of consumer surplus. In the most obvious case the analyst wants to find out what people are or would be willing to pay for a good, such as a clean beach in place of a currently polluted one. But there are other cases: the WTP to prevent a loss or some bad event, the *willingness-to-accept-compensation* (WTA) to tolerate a loss, and the WTA for foregoing a benefit. In all cases we are dealing with a hypothetical situation, and the procedure involves asking people questions. Contingent valuation asks people what they are willing to pay or to accept, while contingent ranking asks them to link their preferences for an environmental good to their desire for other goods that do have market prices. Both techniques are a form of market research, and have grown up in relatively developed countries, especially the USA, where expertise in such surveys is readily available and the public are accustomed to them; in countries where this does not hold, the techniques can still be used, but are often restricted to use goods. Contingent ranking is seldom used, and this section concentrates on contingent valuation.[13]

In general, contingent valuation can apply to any kind of good, but it is usually considered only for non-use goods or values that cannot be tackled by any other method. The procedure is to select a number of people (a sample) and ask them carefully worded questions about the value they place on some hypothetical change. For example, 'Would you be willing to pay $65 a year to have the South Beach kept free of pollution?', or 'Would you accept the loss of Birnam Wood (which is in the way of the proposed Dunsinane Bypass) if you were given £42 as compensation?' These are a WTP and a WTA question, respectively; the latter case could also be addressed by a WTP question such as 'Would you be willing to pay £42 to keep Birnam Wood as an environmental asset?' All these examples are yes/no or referendum-type questions, in contrast to open-ended or direct questions such as 'How much would you be willing to pay to see the beach kept clean?' Once the survey has been done, the answers are analysed to find the average WTP or WTA, which is the measure of economic value for the good in question.

A contingent valuation study is, however, beset by problems, and a good one is correspondingly time-consuming and expensive. The results are very sensitive to biases, both conscious and unconscious. For example, respondents may be confused or misled about the nature, extent, timing or reliability of the change they are being asked about, they may confuse a local effect with a global one (some great crested newts with all great crested newts), they may be influenced by the way they think the hypothetical payment will or might be paid, and they may answer tactically in the hope that they will get some benefit at the expense of others (the free rider problem). They may give an extreme response as a protest or to express an ethical view. The choice of the sample may be critical. The result may be affected by the way the survey is done (face-to-face interviews, telephone interviews, questionnaires sent by mail or email, etc.). The analysis of results is complicated and difficult.

In one case in northern England, the choice of route for a bypass road depended on the value placed on preserving a wood.[14] A contingent valuation survey was conducted, using both WTP and WTA approaches. A thousand questionnaires were sent out to local residents, but only about a third were returned. Many of those were protest responses, generally indicating that protesting respondents did not accept the use of money, or any finite amount of money, to put a value on the preservation of the wood. Quantifiable responses, from the remaining small fraction of the people asked, ranged widely. The results were inconclusive, as the indicated decision between the routes depended on what population was considered to be relevant and represented by the respondents, whether the analysis excluded one outlying response that fell more than 10 standard deviations from the mean, and which of the two surveys was used (the WTA one gave an estimate of value more than 10 000 times bigger than the WTP one). In this case, at least, contingent valuation did not help with decision-making.

Guidelines are available to help analysts do good contingent valuation studies. Generally yes/no questions are better than open-ended ones and face-to-face interviews (expensive), or telephone interviews are better than mail surveys. Questionnaires should be very carefully prepared and tested. The information given to respondents should be carefully thought out and standardised. The sample design and data analysis should use proper and thorough statistical methods, preferably permitting cross-checks. Finally, the analysis should be fully and frankly reported and the full data set should be kept available for re-examination.[15]

The main drawback of contingent valuation is that it depends on what people say rather than on what they do, and the consequences can only be partially mitigated by good methodology. In practice, WTP and WTA measures usually produce widely differing values for the same good, often by a factor of two to five, despite economic theory that claims they should converge if the method is good. It is generally considered that the WTP approach is the better one.

Contingent valuation should be carried out and interpreted by experienced specialists with the appropriate statistical and survey skills, and decision-makers should be well briefed on its limitations. Wherever possible it should be cross-checked against other methods.

C.5 Proxy markets and other revealed-preference approaches

There are some environmental impacts which, although they are not themselves bought and sold in any market, are related to other things that do have a market price. In such cases one can use those linkages to arrive at imputed economic prices for the environmental goods, and this may be easier and/or more accurate than contingent valuation with all its problems. This section describes some of the analytical methods available; readers of this book are unlikely to undertake such analyses themselves, but should know enough about the methods to discuss them with specialists and explain them to decision-makers. The sub-sections that follow introduce the names or titles that are likely to be met with, but the terminology is not standardised and some people regard some of these categories as subsets of others.[16]

Travel cost

This is a method suitable for valuing environmental or recreational sites or facilities when it cannot effectively be done by reference to entrance charges (either there is no charge, or it is suspected to be much less than the whole value). The principle is that users or visitors to such a site reveal the value they place on it by how much time and money they spend travelling to it. The usual procedure is to divide the visitors, or the places that they come from, into groups or zones, and then plot the number of visits made against the cost of travelling to the site, plus entry fees and equipment costs, if any. The area under the graph (the summation of costs for all visits) gives a measure of the economic value that society places on the site or facility.[17] Usually the cost includes the value of visitors' time when travelling, by some approximation.

The travel cost method has the advantage that it relies on people's actions rather than mere words, but there are several problems and difficulties. It measures use values only. People may enjoy the journey to the site, so counting the travel time entirely as a cost may be wrong. People may make the journey for several purposes, not just to see the site, so it may be wrong to attribute the whole of their travel cost to the site's value. Sampling and statistical methods may introduce bias. The method works best for isolated sites (because travel costs are significant and multipurpose journeys are not a problem), and has been applied mostly in rich countries, especially the USA where it was developed.

Avertive behaviour and defensive expenditure

People tend to react to the threatened or potential loss of a good by taking steps to avert or avoid the loss, or to acquire a substitute for the lost good; this behaviour can provide a measure of the good's value. The method can be suitable for valuing pollution, risks of erosion

185

or flooding, or other sorts of land degradation. It works where people are aware of the threats and take actions whose costs can be estimated. Avertive expenditure and travel cost methods are sometimes grouped together under the name *household production function.*

The kinds of avertive or defensive behaviour include direct measures to prevent deterioration (e.g. erosion control measures or water filters), purchase of surrogates (e.g. tanker or bottled water), and relocation to avoid environmental deterioration.

Problems of this method include the fact that some sufferers may have relocated before the survey, leading to an underestimate of the value of an environmental degradation or loss. Substitutes or surrogates may replace only part of the lost value. People may not be well enough informed for their behaviour to be a good guide to true value.

Hedonic pricing

Hedonism concerns pleasure, and this method relies on differentials in market prices to show what value people put on the pleasure of some desired environmental feature, or on the displeasure of putting up with an undesirable feature such as noise or air pollution.[18] The most common application is to property prices, but wage differentials are sometimes used (property and labour are then termed *surrogate goods*). Property prices (either houses or land) can reveal the value of things pertaining to location, such as being near a park, downwind of a smelly factory or facing the sea. Wage differentials can also show location effects, or can help to value unhealthy or unpleasant working conditions. In either case the analysis has to make allowance for other factors such as type of house or qualifications of workers: it is quite difficult to strip out extraneous factors so as to be left only with the differentials that pertain to the good being studied. Hedonic methods have been used particularly for air and water pollution and aircraft noise in developed countries, and for dangerous working conditions.

Hedonic pricing methods require a great deal of data and very sophisticated statistical techniques to disentangle the many factors affecting property prices and wages. The environmental effect or good being valued should preferably be measurable (e.g. aircraft noise) and must be well understood by the people affected, and the property or wage market needs to be reasonably free and efficient (people need to be free to move house or job). The method does use real market behaviour rather than merely stated preferences, but it does not capture non-use values, and its application is very restricted.

Human capital: education and training

Education and training are obviously forms of investment: resources are used up now in the expectation of gaining benefits later. The educated state of a person embodies that investment. By studying the earnings differentials between people with different types or extents of education and training, economists can quantify this investment, at least in a financial sense, and apportion the benefits between the trained individual (private benefits), the government (public benefits) and the whole society, which includes both those parties (social benefits).[19] Earnings differentials are thus used both to value pleasure and displeasure (one of the types of hedonic pricing described above) and to value education and training.

Human capital: health and safety

The name *human capital* is also sometimes given to methods that evaluate the cost of bad health resulting from environmental effects, or risks related to a project or policy.[20] It is consciously limited to the loss of economic production due to a worker's ill-health, and excludes pain, suffering and loss of pleasure that do not happen to be reflected in lost production. Future lost production is discounted back to a base year, and one of the problems of the method is that it naturally puts little or no value on effects on old people. The method is only satisfactory when medical information on the linkages (such as dose–response measures) is readily available. This, together with the fact that only part of the total value is measured, results in very limited application for environmental studies. A related technique is to use lost production in the attempt to put a value on a human life, especially in the analysis of road projects, where the reduction of accidents is a significant objective.[21] In this case a distinction should be made between gross future output and net future output; for the latter, the victim's future consumption is deducted from her future gross output. For valuation, the net future output is normally the relevant one, and it may be much smaller than the gross figure.

Replacement costs and shadow projects

Here the valuation depends on what someone would spend, or would need to spend, to put right or replace something lost, or to counteract the effect or impact being valued. Data come either from observing what people actually do (e.g. raising flood banks to counter river siltation) or from the opinions and estimates of experts who estimate what people would have to do. A shadow project is something under-taken (perhaps only hypothetically) to replace a loss elsewhere; if a road destroys a woodland area of importance as a biological habitat, an equivalent habitat might be deliberately created elsewhere to replace it. In principle, and especially in the case of shadow projects, the method can capture non-use as well as use values, while still using market prices. Analysis requires care over timing and true equivalence; a natural habitat may be able to be effectively replaced elsewhere (although not instantly), but a historical site may not. A special kind of shadow project is afforestation to counteract the effect of a thermal power station on the carbon dioxide balance.[22]

C.6   Shadow pricing as an instrument of policy

In classical free-market theory, distortion of the market is the only acceptable reason for economic prices to diverge from financial prices. Some people, however, recognise other reasons, such as deliberate policy to favour certain sorts of inputs or outputs (e.g. unskilled labour in a place with high unemployment, which is used as an example in Section 2.5). In effect, this sort of shadow pricing is a policy instru-ment used by a government. If the government is truly representative of the people, it is acting as a channel for revealing preferences, along-side the other channels represented by markets and opinion surveys. If it is an unrepresentative government, it is merely applying its own prefer-ences, which may or may not be good (an unrepresentative government can be benign or malign); CBA is a tool and can be used for many purposes.

C.7   Choice of valuation method

The various valuation methods should be regarded as a toolkit, each tool being applied with common sense, where and when it seems the best for the job in hand. An analyst who by habit or ignorance favours one tool over others, or who carries a limited toolkit, is as suspect as a carpenter who does the same. It must be remembered that some methods apply only to some kinds of value (see Box C1), and that the different kinds can sometimes be added (e.g. use value plus existence value). The analyst needs to guard against omission and double-counting, and against false assumptions about cause and effect. A good safeguard is to state the reasoning clearly for others to check and discuss.

Particular problems and advantages of individual methods have been mentioned above. A few generalisations are possible, at least within the field of environmental impact valuation. Methods that are often used, for a variety of reasons, including purely pragmatic ones, are those that rely on quantifiable effects on the production of some easily valued commodity (dose–response functions) and on prevention or replacement (avertive behaviour). Others, such as contingent valuation, hedonic pricing, travel cost and human capital methods, are less often used, largely because they tend to need a great deal of data and the use of sophisticated statistical techniques.[23]

When it is possible within the constraints of data and time to make two independent estimates of the same quantity, it should be done. The record of past studies where this has been done gives a salutary warning about the low level of precision in the valuation of many aspects, especially environmental ones. A factor of three between higher and lower estimates of the same thing is not unusual, nor cause for embarrassment; even with such a range it is often worthwhile to do the analysis and indicate at least the order of magnitude. Sensitivity tests are useful, as always, to show how much the imprecision matters within the context of a whole CBA. Where two different methods give widely differing estimates, it may be possible to choose between them by making the analysis, at least with respect to that parameter, a lower-bound one.

An interesting example of the use of several methods compared against each other is the difficult problem of putting a value on a statistical life, for CBA of projects which produce or avoid accident risk. (Although some people consider it improper even to try to put a monetary or economic value on a human life, decision-makers cannot avoid it and should preferably do it transparently.[24]) Methods used include contingent valuation, hedonic pricing (mainly wage differentials), insurance and human capital approaches. The range of values obtained is wide, reflecting the conceptual difficulty; for the UK alone published results range from less than £250000 to several million pounds (1987 prices), with a good case for a value over £2 million, but in 2010 the British recommendation remains at £1.1 million at 2000 prices.[25]

C.8   Benefit transfer

It is possible to take estimates made in one place and use them in another, which is called 'benefit transfer'. Because CBA is often done with limited time and resources, this is sometimes the only possible way to proceed, if certain classes of costs and benefits are to be included at all. It must, of course, be done with great care, ensuring that the two situations are really comparable. Many valuations are heavily dependent on the nature and state of the economy of the country in which they are done, and should not be hastily transferred to another country merely by the use of currency exchange rates. When the original valuation involved technical linkages such as dose–response ratios, it may be possible to use the technical ratios from a distant place but then apply local conditions to the economic valuation stage.

For example, Pearce (1995) reports estimates from various studies of the cost of air pollution by particulate matter. Seven studies in four continents give results ranging from US$20 to US$1000 per head of population, the variation being largely due to different assumptions regarding the value of a statistical life (from about US$50 000 to US$2–5 million).

C.9   Further reading and information sources

Many readers of this book will obtain all they need on the valuation of environmental costs and benefits from the relatively brief book OECD (1995) (written by Winpenny), and those needing deeper discussion will miss little by restricting themselves to Hanley and Spash (1993) and Winpenny (1991) (full references are given in Appendix K). Pearce (1993) concludes with a 13-page appendix summarising the methods and their applicability. Dixon *et al.* (1994) give brief descriptions of methods and extensive case studies in four continents. Among the more general CBA books, Perkins (1994) includes thorough coverage of most aspects relevant to this appendix, and Schofield (1987) covers most of the methods from a planning viewpoint.

Other books that may be helpful for details include Pearce and Turner (1990), van Pelt (1993), James (1994) and Field (1994). Detailed treatment of some specialised topics is given in Layard and Glaister (1994). Coker and Richards (1992) and Willis and Corkindale (1995) provide challenging collections of papers of interest to readers who want to think about the basis of valuation, especially, but not only, for environmental aspects. Abelson (1996) describes the environmental methods and gives several case studies. Curry and Weiss (1993, 2000) is particularly useful on traded and non-traded goods, labour, and land in developing countries. Bateman *et al.* (1993) discuss valuation methods for environmental impacts of transport projects. Merrett (1977) covers water resources development, and Penman (1999) covers dams.

190

Books issued by or associated with the World Bank, especially the one by Ward and Deren (1991), tend to give a markedly economist-styled approach, based on free-market principles. For the British public sector HMSO (1991) covers many fields; paragraphs 18 and thereafter cover briefly the use of market prices (cleanup, avoidance, replacement, substitute) and revealed preferences (related markets, house prices, travel cost), and emphasise the approximate nature of the results. H.M. Treasury (2003) updates the British context. The USA's environmental issues are covered particularly by Field (1994). A number of references for specific fields are given in Chapter 4 of this book.

Notes

1 The use of human preferences as the basis of value is itself debatable. Some people hold that things can have value independently of the appreciation of humans or other sentient beings. A pragmatic view is that value is independent of humans and their perceptions, and human preferences are just a measure of value, not a perfect one but the only practicable and accessible one. (A badger's appreciation of her sett may have value independently of any human, but it is difficult to ask her to quantify that value. The rocks of the Grand Canyon or the genetic information of the smallpox virus may have intrinsic or existence value, but it is not easy to value them without recourse to human judgement.) In any case, the use of a monetary unit to count values does not mean that life, beauty and pain are being reduced to money, merely that money measures many human preferences and is therefore the handiest unit for counting the others as well. For a discussion of this see Coker and Richards (1992; the paper by Holland and Cox).

2 The various approaches can be classified in a number of ways, which in the literature are usually presented in the context of valuing environmental aspects. Field (1994) and van Pelt (1993, 1994) use a distinction between cost and benefit aspects. Winpenny (1991, and the OECD 1995 guide which he also wrote), and Hanley and Spash (1993) treat the same methods in different orders. Dixon *et al.* (1994) divide them into objective and subjective approaches. The Asian Development Bank (1997) gives useful guidance, and Penman (1999, Section 3.2.1) has a useful 2×2 classification table. Some people distinguish primarily between use and non-use benefits. The present book follows a classification of methods rather than of benefit or cost types.

3 A rigorous definition and explanation of consumer surplus can be found in most economics textbooks and some CBA books, notably Perkins (1994, Section 7.4), Potts (2002) and Brent (2006, Ch. 3), as well as in Belli *et al.* (2001).

4 The explanation of this is outside the scope of this book; it can be found in an economics text such as Pearce (1983, Ch. 3); it is explained for non-economists in Field (1994, Ch. 3).

5 Ward and Deren (1991) acknowledge that governments may sometimes have good reason to fail in this sense, but regard such cases as rare exceptions to a general rule. Market failure is explained in Perkins (1994, Part II).

6 Guidance for economists on valuing particular kinds of costs and benefits is in the specialised chapters of many books cited here, and in the guidelines of particular institutions, such as the Asian Development Bank and the World Bank.

7 Pragmatic discussions of the economic valuation of labour are given in Gittinger (1982, pp. 258–263) and in ODA (1988, Ch. 3). Both of these use the foreign exchange numéraire, but the principles can also be applied using the domestic price one. A thorough treatment is given in Perkins (1994, Ch. 10). For urban unskilled labour from a rural source, see Brent (1990, 1998). Curry and Weiss (1993, 2000) and Potts (2002) discuss the valuation of labour in developing countries under various conditions and both numéraires. Shadow pricing of labour is often appropriate in developed countries too; the journal *Impact Assessment and Project Appraisal* carries relevant papers.

8 If the land would be taken permanently the appropriate discounting period would be perpetuity; the PV/RV factor for very large or infinite values of n is simply $1/r$; for instance, 20 when $r = 5\%$ and 12.5 when $r = 8\%$. Some economists prefer to use for this discounting, instead of the normal discount rate, the opportunity cost of capital minus the expected real growth rate of gross domestic product; this seems inconsistent with the discounting for other purposes in the same analysis, although it generally makes little practical difference. Curry and Weiss (1993, 2000) explain the valuation of land using both numéraires.

9 The Asian Development Bank's guidelines are particularly thorough on land (Asian Development Bank, 1997, Appendix 13).

10 Ward and Deren (1991) give details and examples on valuing traded and non-traded goods (generally using the foreign exchange numéraire, but the logic of adjusting for transport costs is independent of that). Their case study 7 shows the large difference that transport costs can make between import and export parity prices, and how a weighted mean can be used when neither of them applies all the time. Curry and Weiss (1993, 2000) explain the valuation of traded and non-traded goods with both numéraires. See also Perkins (1994, Ch. 8).

11 Ward and Deren (1991, Ch. 12, Technical Note 23) give examples; they use the foreign exchange numéraire, but the logic applies with any numéraire.

12 A practical account of contingent valuation is given in OECD (1995, Ch. 6) or in Green *et al.* (1991), while a more theoretical approach is

given in Hanley and Spash (1993, Ch. 3). Dixon *et al.* (1994) describe contingent valuation in the context of projects supported by the Asian Development Bank and the World Bank.

13 Contingent ranking is described in Hanley and Spash (1993, Appendix 3.1).

14 This is the Harrogate–Knaresborough Bypass, as described by Hanley and Spash (1993, Section 12.3).

15 A thorough review of the method (NOAA, 1993) was made after the *Exxon Valdez* oil spill in Alaska, and its recommendations are set out in OECD (1995, Section 6.4) and Hanley and Spash (1993, Section 3.6). Advice on contingent valuation can also be found in or via the British government's Green Book (H.M. Treasury, 2003).

16 Most of the methods described here come under the title 'market valuation of physical effects' in OECD (1995, Ch. 5). Advice can also be found in or via the British government's Green Book (H.M. Treasury, 2003).

17 Most textbooks refer to the summed travel cost as 'consumer surplus', although the consumer has actually paid at least part of the cost. Many books also exclude the entry fee. A thorough treatment is given in Hanley and Spash (1993, Ch. 5), and Zerbe and Bellas (2006) give an example.

18 Hedonic pricing is described in OECD (1995, Section 7.3), Hanley and Spash (1993, Ch. 4) and Winpenny (1991, Section 3.3.4).

19 This sort of human capital analysis is described in Johnes (1993, especially Ch. 2). The distribution of benefits between the parties is a particular theme of Curtin (1996). See also Section 4.5 of this book, and its notes.

20 Human capital, in this sense, is discussed in Winpenny (1991, Section 3.3.3) and Hanley and Spash (1993, Section 6.2.2).

21 For fatal accidents the valuation of lost output can be compared with values derived from two other approaches: insurance premiums and court settlements. All are discussed briefly in Adler (1987, p. 41).

22 This is discussed by Winpenny (in Weiss, 1994a, Ch. 8).

23 Comments on the relative merits and applicability can be found at the beginning and end of each section in OECD (1995, Chs 5–7) and in Winpenny (1991, Section 3.3.7). The application examples given in Winpenny (1991, Ch. 4) also give practical guidance. A theoretical comparison is given in Hanley and Spash (1993, Ch. 7).

24 Any decision between two courses of action involving different costs and different risks of loss of life imply a value for a statistical life, whether the decision-maker realises it or not. Analysis of such decisions can reveal the implied values, and they tend to vary very widely (Jones-Lee, 1989; Belli *et al.*, 2001, especially Chs 6 and 9).

25 Jones-Lee (1989), Adler (1987), COBA (1996 and updates). In the British context, H.M. Treasury (2003) discusses the 'value of a prevented fatality',

quoting £1.1 million at 2000 prices as a figure to be multiplied by the estimated number of deaths on roads. Another guideline gives the same figure for deaths by flooding incurred or avoided (supplementary note, May 2008, to Environment Agency (2010)).

D

Distribution of costs and benefits between people

D.1   General distribution adjustments

In its simple and usual form, economic CBA regards a unit of benefit or cost as equally valuable, regardless of who experiences it. Decision-makers may, however, want also to consider the distribution of benefits and costs, as well as their totals. For instance, they may wish to give greater weight to some people than to others; the people can be grouped by time, location, gender, wealth, or some functional description such as coalminers or hill farmers.

Ways of using distributional considerations in guiding a decision include:

(a) **Independent description**: distributional factors are examined and reported alongside the economic indicators of the CBA, without any attempt to combine them or evaluate trade-offs.

(b) The **distributional-constraint approach**: economic efficiency is maximised subject to a constraint on distributional effects. Any project not meeting the threshold of distributional equity or accept-ability is rejected, and of the rest the most economically attractive is preferred, regardless of relative distributional effect.

(c) Its inverse, the **economic-constraint approach**: all projects that fail to meet a minimum economic standard, such as a cut-off IRR, are rejected, and of the rest the one that shows the best distributional effects is chosen, regardless of its economic indicators.

(d) The *multicriterion analysis approach*: the indicator of economic efficiency (normally measured by a distribution-neutral economic CBA) is placed alongside one or more distributional indicators in a multicriterion analysis, as discussed in Appendix F, the indicators being treated as independent. Trade-offs between efficiency and equity can then be handled by explicit and transparent weights in the multicriterion framework. (The constraint approaches (b) and

(c) could, of course, be regarded as special cases of this multi-criterion one, as multicriterion methods can easily accommodate threshold constraints.)

(e) The **weighted CBA approach**: costs and benefits are weighted according to who experiences them, and the CBA proceeds with this weighting incorporated within it, just like any other weighting or adjustment such as shadow pricing. (CBA with explicit weighting is sometimes called 'social CBA' as opposed to 'economic CBA', but this term tends to cause confusion because the label *social* is also used to distinguish collective from individual preferences, as in Appendix E.)

In practice, distribution aspects are seldom included in a CBA, despite being discussed in many books on the subject. In particular, the World Bank handbook pays much attention to 'who gets what' and has a chapter on gainers and losers.[1] The aim of this brief appendix is to acquaint the reader with the basic ideas and terminology and to indicate where further discussion can be found.

Within CBA, relative weight between people of different times or generations is normally dealt with by discounting, and the implications are discussed in Appendix E.

For weighting between contemporaries within CBA, the weighting factors can be directly and subjectively defined by the decision-makers, for instance if they specify that all costs and benefits affecting tin miners in the Misty Mountains shall be weighted by a factor of 1.3 relative to everyone else.[2] This sort of weighting is transparent and subjective. It can mix wealth/income criteria with regional or gender ones, or any other kind of criterion; in the unlikely extreme, for example, special weighting might be desired for poor, elderly, or ethnic-minority males dependent on fishing. It is seldom used, however, perhaps because it is more obviously subjective than other parts of the CBA toolkit.

D.2   Income weighting

A more restricted class of weighting factor depends solely on income (or consumption level), and this class has received much attention in the literature because it can be mathematically formulated and fits comfortably with economic theory. The idea behind income or consumption weighting is that rich people appreciate or value a certain unit of cost or benefit less than poor people do. In financial terms, an extra £1000 of annual income is more significant to someone now earning £10 000 a year than to someone earning £50 000 a year. The same principle

---

Box D1 *Simple income distribution weighting*

The simplest and commonest way to apply a weighting scheme is to assume that if some group of people have a per capita income of P in contrast to the national average per capita income A, then all costs or benefits affecting that group are multiplied by a weighting factor d, given by

$$d = (A/P)^n$$

where n is an index representing the degree to which the decision-makers want to apply income weighting. If n is zero, all weights d are 1.0 and there is no weighting at all. If n is 1 the weight is inversely proportional to income P. An n value of 0.5 represents fairly weak weighting and a value of 2 is strong (with things that affect people on half the national average income being given four times the weight of things that affect average earners). The values suggested for actual use usually range from 0.5 to 1.5.

---

applies in non-financial senses, for instance to environmental goods. Box D1 summarises the basis of the standard procedure.[3]

The income weighting scheme in Box D1 is a smooth function of income because of the constant index n. An alternative is the basic needs approach. This uses a kinked weight/income function, recognising thresholds of poverty or wealth. For instance, everyone above a certain threshold income may have the same weight regardless of how rich they are, and those below the threshold ('the poor') may have a considerably higher weight, increasing rapidly with decreasing income below the threshold. This is a more targeted weighting scheme and avoids giving very low weight to large sections of the population.[4]

Any weighting system is subjective and heavily laden with value judgements. What is not always realised is that the decision not to use a weighting system is equally so: giving equal weight to all groups is just one weighting system among many, although it does of course happen to be the simplest to apply. In economic theory the omission of all explicit weighting is associated with economic efficiency and the acceptance of the prevailing income distribution, but a decision to apply the theory of economic efficiency is itself value-laden.

An argument for keeping distributional weights out of CBA is that if the prevailing income distribution is not optimal there are other ways of correcting it (e.g. taxation), and CBA is not a good instrument for the purpose.[5] The counter-argument is that, in many countries, tax and welfare systems do not serve to reduce inequality (for political reasons), so that implementing infrastructure or development projects that favour the poor *is* an appropriate, feasible and desirable means to that end. Some economists argue that, if economic efficiency must be sacrificed for distributional purposes, the degree of sacrifice should at least be quantified and kept below some acceptable limit.

If and when distributional weighting is done, the choice of weights (or of the index *n* in Box D1, which largely determines them) is difficult and controversial. If weighting turns out to be crucial for a decision, the analyst should carry out sensitivity tests on the weights (e.g. finding the switching value of a significant weight).[6]

D.3 Use of distributional weighting

Historically, the application of mathematically derived income weights within CBA was advocated by several authors of guidelines in the late 1960s and 1970s, but was seldom actually done. More recent guidelines tend either to ignore the issue altogether or to advocate reporting of distributional effects alongside a CBA without trying to incorporate them within it.[7] The latter approach corresponds to (a) in Section D.1 and is recommended for most situations. A CBA that starts by collecting data in financial form, distinguishing between parties such as the government, the project authority and the beneficiaries, gives a good basis for presenting the group-specific financial impacts alongside the national economic CBA.

An intermediate possibility is to treat distributional issues alongside CBA in a multicriterion framework, as indicated in approaches (b), (c) and (d) above.[8]

The incorporation of distribution weights inside CBA, approach (e), is only likely to be appropriate in rare special cases when the decision-makers clearly understand it and explicitly desire it. In any other circumstances it is likely to be insufficiently transparent, even if it is declared in the report.

Notes

1 Belli *et al.* (2001, and its earlier drafts). Potts (2002) also covers distribution effects and poverty reduction.
2 This direct setting of weighting factors is discussed in van der Pelt (1994, Section 5.6.2).

3 This is a simplified presentation of the theory. More strictly, the weights are usually derived from consumption levels rather than income. The index n is usually identified with the elasticity of the marginal utility of income or of consumption (implying that the elasticity is constant), but it can be regarded just as a parameter defining the weighting function. Formal presentations, with slight differences between them, can be found in Irvin (1978, Section 8.04), Pearce (1983, Ch. 5), Brent (1990, 1998), Curry and Weiss (1993, 2000) and Perkins (1994, Ch. 14).

4 The basic needs approach is discussed in some detail in Brent (1990, Chs 12, 13) and in Ray (1984).

5 See Layard and Glaister (1994, p. 47).

6 Perkins (1994, Section 14.2).

7 Examples are the World Bank handbook (Belli *et al.*, 2001), which emphasises description of distribution (gainers and losers) but does not generally advocate any weighting, and Winpenny (1991, Section 3.4.2). A systematic way of presenting distributional effects, used in the USA water sector, is called multi-objective display matrix, and is described in Schofield (1987, Section 6.3). Weighting is also discussed in Potts (2002) and Penman (1999).

8 This is discussed particularly by Schofield (1987, Ch. 6) and van Pelt (1993, Sections 3.2.3, 6.6, 6.7).

E

Choice of discount rate

E.1   Introduction

The process of discounting is introduced and described in Chapter 3, starting from the concept of an individual's time preference rate and then assuming a collective time preference rate for a whole society or nation or other group of people on whose behalf a CBA is being conducted. Precisely why we should discount the future, i.e. have a positive time preference rate, is not easy to explain and justify, and there are many different approaches. This appendix explains a few of them and points the reader to more specialised texts: it is intended to amplify the explanation in Section 3.9, and for completeness and clarity includes some duplication and overlap with that section. If the reader is, like the author, a non-economist, he will find most of the published works cited in this appendix difficult to understand, as they are couched in specialised economic language and use graphs, formulae and mutually inconsistent abbreviations. This appendix should, however, enable him to recognise some of the strands and tendencies, and to derive some illumination from the specialised books and articles, even when they cannot be followed in detail.

The discount rate discussed here refers to computations done in constant-value terms (see Section 3.5 and Box 3.6, page 70), and has nothing to do with inflation.

The standard practice of discounting makes several assumptions, which are seldom challenged but which are not self-evidently true, such as that the rate should be the same throughout the analysis period, and that it is right to apply the same rate to all different kinds of costs and benefits. Some of the published literature quoted here discusses alternatives to these assumptions, but they are almost always applied in practice.

The test or threshold discount rate used to guide decisions is related to the balance between consumption, on the one hand, and savings and investment, on the other, because a high test discount rate will tend to reject more investment projects than a low one. As well as the balance between any individual's consumption now and his consumption later (by way of savings and investment now), the discount rate concerns intergenerational balance and equity.

In general both the underlying rationale and the practical guidance on choosing a rate fall under three headings:

- time preference as an observable human characteristic
- the opportunity cost of resources, especially of capital
- the capital rationing device.

These are now discussed in turn, and then Section E.5 discusses the involvement of risk and uncertainty, and Section E.6 provides a summary and discussion of rates commonly used. The last section lists some references to the literature.

E.2 Individual and social time preference

Individual people, and even laboratory rats, can be observed to show a preference for good things to happen sooner rather than later, and bad things later rather than sooner: by implication they therefore apply a positive discount rate. (It is fashionable at this point to quote proverbs such as 'a bird in the hand is worth two in the bush', 'mas vale pajaro en mano que cien volando',[1] and 'un tiens vaut mieux que deux que tu l'auras'.[2] These illustrate a widespread and deep-rooted time preference, but they also show that most people (except apparently the French) tend to confuse timing with probability of access (see Section E.5). The apparent north–south ratio gradient, from two to a hundred across the Pyrenees or between the Americas, is not statistically well-founded.)

In the case of rational individuals making individual decisions, this so-called marginal rate of time preference has several reasons behind it, notably:

(a) impatience (partly irrational, sometimes combined with the thought that there is a finite probability that the individual will not be alive to enjoy a future benefit – sometimes called pure time preference rate or myopia)
(b) the facts that many people are getting richer in real terms and that rich people appreciate the marginal unit of goods less than poor people do (called, respectively, the *rate of growth of per capita consumption* and the *diminishing marginal utility of consumption*).

Thus far the individual's position. When many individuals act in a free market, each deciding when and how much to consume, to save and to invest, their behaviour can, in principle, make evident what their individual time preference rates are. Insofar as CBA aims to apply the wishes of the people, these revealed individual rates provide a starting point for arguing what the collective or social rate ought to be (note that 'ought to be' is a normative statement, embodying value judgements, and is thus outside the narrow realm of so-called 'positive' economics, which claims to be value-free). In this context the word *social* means collective rather than individual, and has little to do with the social/economic distinction made in Section 1.3, although that also implies value judgements.

The *social time preference rate* (sometimes called the *consumption rate of interest*), which is the rate which the people wish to see applied to collective decisions, is, however, likely to be lower than the marginal time preference rate they apply as individuals. This is because they are aware of goods that no market can provide, and of the ethical rights of future generations. How far below is infinitely debatable, but it is widely agreed that the agents acting for people at the collective level (i.e. governments) have a responsibility firstly to reflect this difference between individual rates and what individuals wish for collective rates, and secondly (debatably) to go beyond that and represent the interests of future generations. This last step can be seen as paternalism overriding democracy, but may nevertheless be ethically justified.[3] In game theory it can be argued that, in the case of decisions about investing for the future

> *everyone is made better off by having the decision made with reference to the social discount rate rather than the individual discount rate. The social discount rate is lower than the individual discount rate, and it is reflected only in public decisions.*[4]

There are indications that the social time preference rate is often in the range 2–6%, although it cannot be measured without controversy. The British Government, basing its discount rate on the social time preference rate, recommended 6% in the 1990s, but since 2003 this has been 3.5%, and even less for long-term projects. Work for the World Commission on Dams suggested social time preference rates as low as 0.5–3%.[5]

E.3   Opportunity cost of capital

Applying the opportunity cost principle to discounting, one can argue that the discount rate should reflect what society foregoes elsewhere if it uses certain resources for a period of time in a certain project. In traditional economic theory the measure of this is the real interest rate or

> *return on the last or marginal investment . . . that uses up the last of the available capital. If set perfectly, the rate would reflect the choice made by society as a whole between present and future returns, and, hence, the amount of total income the society is willing to save.*[6]

This real interest rate, which is often used as a guide to the appropriate discount rate, is called the *social opportunity cost* or the *opportunity cost of capital*.[7] It can be regarded as an observation of the investment behaviour of producers, giving an indication of their time preference rate, just as observation of consumers reveals the individual time preference rate.[8]

The actual determination of the opportunity cost of capital is a job for professional economists (and they seldom agree about it); they usually analyse the real interest rates (see Box 3.6, page 70) and rates of return on investments in different categories and sectors. Typical rates actually used are in the range 8–15%, with a range of 10–12% commonly used in less developed economies.

Some analysts and economists regard the opportunity cost of capital as the main, or even the only, correct guide to the right discount rate to use for guiding decisions, but others disagree. The interest rate prevailing in a market subject to corporate and individual taxes can be expected always to be higher than the true rate of return on investments.[9] This means that the private sector will undertake too little investment from society's point of view, failing in particular to invest in long-term projects.[10] Many argue that the government, as the representative of collective interests, should undertake investments that the private sector does not find attractive. This again implies that collective decisions (involving economic and social CBAs in the sense of Section 1.3) should be guided by a lower discount rate than private ones.

E.4   Capital rationing device

The third aspect of the discount rate relates to its practical role as a capital rationing device in a financial sense, or a resource rationing device in a more general sense. If a particular country or sector has a limited amount of resources in a certain year or other time period, and the resources are allocated to projects by judging each project's CBA indicators against a test discount rate or limiting internal rate of return (IRR), then the higher the test rate the fewer projects will be approved and implemented. If the rate is too high there will not be enough projects being implemented to use up the resources, and if it is too low the resources will run out. In principle, there is an ideal rate somewhere in between, which will just achieve the full use of resources.

In contrast to the other two approaches, which try to work out what the discount rate ought to be according to some theory or set of ideals, this is a pragmatic approach that merely accepts what the discount rate actually does in practice (i.e. regulate the stream of projects that get implemented) and looks for a rate that works. The approach is useful when there really is a definite limited resource, in which case the only problem is to find out what the test rate should be to keep things in balance in the medium or long term, without being distracted or misled by short-term fluctuations in the supply of either resources or eligible projects. However, often the resource is not so neatly bounded, nor is the class of projects that compete for it. Resources are often not available for just any use, but are earmarked for some particular kind of project. The discount rate arrived at by this approach is only as appropriate as the supply of resources, and that supply may, for a particular purpose or category, be far from optimal; it does not seem very satisfactory, for instance, to let the discount rate for a country's road projects depend on the amount of loan funds the World Bank chooses to make available for that sector. Sometimes a CBA needs to give an indication of a project's inherent economic or social merit, independently of whether or when a country can afford it or finance it.

It is therefore difficult to use this approach by itself to determine an appropriate discount rate for a country, but the concept of a rationing device is useful for judging the performance over time of a system of resource allocation that uses CBA and discount rates as one of its decision-guiding methods. In the limit, over a whole economy, the capital-rationing principle tends to merge with the opportunity cost approach.[11]

E.5 Risk and the discount rate

In financial markets the interest rates and returns on investments are related to the degree of uncertainty and risk associated with particular projects or sectors: people sometimes talk of a *risk premium* in the sense of an addition to normal rates of return, to allow for uncertainty. Just as interest on any loan or investment is seen as a reward for waiting, the extra interest on a risky one is seen as a reward for risk-taking. This is a sensible and useful concept for investors, who see a trade-off between risk and return, and consciously balance their portfolio along that spectrum.

A similar concept is sometimes applied to discount rates for CBA, partly because of the link between interest and discount rates by way of the opportunity cost approach. Some people therefore consciously add in a risk premium in arriving at a discount rate for CBA. This is, however, generally not advisable, notably because:

- it confuses what ought to be two distinct matters, namely risk on the one hand and the time value of resources on the other
- it assumes that riskiness increases with time and can be allowed for by heavier discounting of far-future costs and benefits, which may not be at all realistic for a particular project's riskiness.[12]

The concepts of risk pooling and risk spreading can be said to imply that a decision for a large group of people should use a lower discount rate than a decision for a smaller group, and hence that public sector CBAs should use lower discount rates than private sector ones. The argument is, however, not recommended, because of the above general drawbacks to the concept of any risk premium in CBA, and because it encourages confusion with other and better reasons for differentials between public and private sector analyses.

It is recommended that, in any kind of CBA, the whole matter of risk and uncertainty should be kept separate from matters of timing and should be dealt with explicitly, preferably by means of sensitivity tests and perhaps probabilistic risk analysis.[13] Appendix A discusses the treatment of uncertainty in CBA, including the arguments on risk pooling and spreading.

E.6   Synthesis

The above summary of the different aspects of and approaches to the choice of discount rate should make it unsurprising that there is a wide range of rates used in practice. Although in economic theory the good allocation of resources, especially capital, requires that the same rate be used in all sectors of a national economy, there tend to be different rates for different purposes. The rate used also depends in practice on the identity of the prospective decision-maker or the audience for each report or analysis.[14]

If the context is dominated by bankers or financiers the rationale of opportunity cost of capital tends to predominate, sometimes accompanied more or less explicitly by capital rationing considerations, and the rate used is usually in the range 8–15%, typically 10–12%.[15] This represents a very short-term view. (If a generation is taken to be

205

30 years, it means that things that happen when our children reach our age have only one-twentieth the weight of things that happen in our time.)

If the context is more one of political decisions taken on behalf of a nation with a long-term view and ethical considerations, the (social) time preference rate approach is more likely, and the discount rate may be as low as 2–3%.[16]

Some kinds of decision and project, especially environmental matters and forestry, have a particularly long-term perspective. People working in these fields find that their projects can seldom be justified on criteria that involve the higher discount rates, but feel that the projects should be implemented for the long-term good of both people and the planet. This has led to a lively debate on the rationale of discount rates, and many of the more recent references in Section E.7 reflect this.[17] The debate involves the rights of future generations, and some ingenious modified discounting schemes have been suggested, such as the one of Kula, which discounts separately for people born at different times.[18] In practice, the effect of special schemes is similar to that of choosing a low discount rate, and in the early 1990s this was recognised by the UK government in sometimes applying a lower rate to forestry projects than to others. Later British practice used lower rates for later years of long-term projects in any sector.[19]

Some people advocate a conscious weighted-average discount rate that takes account of all approaches.[20] Whether this is stated or not, the rates used in practice do often originate from some compromise between approaches. In the more developed economies the approved rates for public sector decisions often lie in the range 3–8%. There is, however, a tendency to vary the rate according to the sector involved, as a conscious political decision to favour some sectors more than the application of a standard rate would. In the USA, the development of major water resources projects has historically used a lower rate than other sorts of public sector projects, and indeed it is said that many of the vast projects of the mid-twentieth century would not be found acceptable at normal discount rates.[21]

To summarise, there is no such thing as the correct discount rate: the choice has ethical implications and is ultimately a political one. Many readers of this book will never have to choose a discount rate for themselves, but will ask others to name a rate for a particular analysis. In general, it should be the decision-makers who set the rate. It is, however, useful for analysts and other professionals to be acquainted with the different arguments and approaches, if only to enable them to detect and question a dubious recommendation from someone with a partial or narrow viewpoint.

E.7   Further reading and information sources

This section lists selected works from the extensive published literature, and comments briefly on their treatment of discount rates. Readers who wish to follow a more up-to-date discussion would do well to look at the monthly journal *Impact Assessment and Project Appraisal*.[22] The full references are given in Appendix K.

- Potts (2002): an economist's rigorous estimation of appropriate discount rate, involving many value judgements.
- Cline (1993): part of a special issue on discount rates; argues for very low rates.
- Curry and Weiss (1993, 2000): main emphasis is on opportunity cost of capital and capital rationing.
- Eckstein and Lecker (1995): examines some unusually shaped financial cash flows and shows how they affect capital rationing and project ranking.
- Hanley and Spash (1993): a clear exposition in Chapter 8 of the reasons why the social time preference rate is lower than the individual one and than the real market return to capital; extensive discussion of the ethics of intergenerational distribution.
- H.M. Treasury (2003): discount rates in Paragraphs 3.51 and 3.63, also Annex 6; recommends 3.5% for most British public-sector decisions, based on a formal estimate of the social time preference rate. For the later years of analysis, i.e. periods longer than 30 years, lower rates are recommended. Discussion and references provided.
- Irvin (1978): discount rate rationale in Sections 1.04 (general), 7.07 (economic) and 8.04–8.07 (social).
- Layard and Glaister (1994): Chapter 3 is a thorough and mathematical treatment of discount rates by J.E. Stiglitz.
- Livingstone and Tribe (1995): discusses the effect of different discount rates on particular sectors, and then summarises the main approaches and quotes typical rates for the UK.
- Markandya and Pearce (1994): a clear summary of the components of social time preference rate, plus a comparison with the opportunity cost of capital, and a commentary on sustainability and environmental concerns.

- ODA (1988): discount rates are treated fairly briefly, mainly in Paragraphs 3.100–3.111.
- Pearce (1983): pages 40–50 give a theoretical economics approach to the contrast between the opportunity cost of capital and the time preference rate.
- Pearce (1993): Chapter 3 covers discounting in the environmental context.
- Pearce (1994): covers the particularly sensitive case of forestry projects.
- Pearce and Turner (1990): Chapter 14 explains the rationale of discounting from all approaches and adds an environmental critique.
- Perkins (1994): Chapter 13 describes the various theoretical approaches.
- Price (1993): thoroughly examines the technical and ethical details of the rationale for discounting and gives many arguments against it, then discusses other ways that allocation decisions might be made without discounting.
- Roberts (1996): a senior economist advising bilateral aid donors argues for the opportunity cost approach.
- Sassone and Schaffer (1978): Chapter 6 reviews the various rationales, distinguishing clearly between personal and collective time preference, and between market and ethical considerations.
- Squire and van der Tak (1975): summarises the approaches associated with the two numéraires and analyses the discount rate in that context.
- van Pelt (1994): discount rates in Section 3.4.7, concentrating on opportunity cost of capital; thorough discussion of discounting for environmental aspects in Section 4.3.3.4.
- Ward and Deren (1991): Section 21 expounds the consumption rate of interest and the accounting rate of interest.
- World Bank (1996): this handbook (in its Technical Appendix) said the Bank had traditionally used 10–12%, as a rationing device:
  > task managers are free to use a higher or lower rate where warranted, as long as they provide a sound justification... a rate lower than 10% might be difficult to justify.

This paragraph has been omitted from the published version (Belli et al., 2001). Both versions give a full account of the conceptual framework for determining the opportunity cost of capital.

- Zerbe and Dively (1994): Chapter 13 treats the rationale of discount rates from the starting point of a world with taxes as a second-best kind of market, and relies heavily on the market in government bonds.

Notes

1 'A bird in the hand is worth more than a hundred flying'.

2 Roughly, 'something you have is worth more than something you will have'.

3 These distinctions are discussed in Markandya and Pearce (1994, p. 33), Hanley and Spash (1993, Section 8.2, pp. 128–129) and Price (1993).

4 Sassone and Schaffer (1978, Section 6.3.2, p. 111). Academic economists' accounts of the social time preference rate can be found in Potts (2002, Ch. 12) and Brent (2006, Ch. 11).

5 H.M. Treasury (2003), Penman (1999).

6 Gittinger (1982, p. 314). See also Irvin (1978, Section 7.07, p. 129).

7 In principle, this social opportunity cost is closely related to the social time preference rate discussed in the previous section (see Irvin 1978, Section 7.07, where the opportunity cost approach is used first. Irvin also makes a clearer distinction than most between economic and social analyses).

8 Hanley and Spash (1993, Section 8.2, p. 128). Pearce and Turner (1990, Section 14.2) say

> The basic observation about capital is that if we divert some resources for investment (capital formation) rather than consumption, those resources will be able to yield a higher level of consumption in a later period than if we consumed them now

and go on to show how the opportunity cost of capital can be derived from a production possibility curve. The latter is also in Pearce (1983, Ch. 4). The thorough estimation of an economy's weighted mean opportunity cost of capital is demonstrated in the Technical Appendix in Belli *et al.* (2001).

9 The argument is set out in some detail, under the title of 'A second-best world', by Zerbe and Dively (1994, Section 13.2); to a neo-classical economist the first-best world has not only perfectly informed consumers in a totally free market but also no taxes.

10 Hanley and Spash (1993, Section 8.2.1).

11 See Curry and Weiss (1993, p. 115) for different sorts of resource limit. Gittinger (1982, p. 314) introduces the opportunity cost of capital as

> the rate that will result in utilisation of all capital in the economy if all possible investments are undertaken that yield as much or more return.

See also Irvin (1978, p. 130) and Eckstein and Lecker (1995). Roberts (1996) uses the capital rationing approach to argue for the opportunity cost of capital.

12 This is explained further in Brent (1990, Section 11.1.2).

13 The risk premium concept is discussed in Irvin (1978, p. 52) and in Hanley and Spash (1993, Appendix 8.1, p. 146). Pearce and Turner (1990, Section 14.3) and Perkins (1994, Ch. 14) give reasons for keeping risk premiums out of CBA.

14 Strictly speaking, the definition of the discount rate is tied up with that of the numéraire or common unit of account (see Section 2.3): the discount

rate reflects the way the value of the numéraire changes over time. The terms *consumption rate of interest* and *accounting rate of interest* are sometimes used in connection with the consumption and investment aspects, respectively (see Squire and van der Tak, 1975, pp. 69, 75). The latter term is explicitly linked to the foreign exchange numéraire. For a more recent statement of the World Bank's attitude, see Ward and Deren (1991, Section 21) or the Technical Appendix in Belli *et al.* (2001).

15 See ODA (1988, Sections 2.8, 3.100–3.111). Gittinger (1982, p. 314) says:

> No-one knows what the opportunity cost of capital really is. In most developing countries, it is assumed to be somewhere between 8 and 15% … a common choice is 12%.

The 1996 and 1998 versions of the World Bank handbook stated candidly that the Bank uses 10–12% as a notional rate, primarily intended as a rationing device for the Bank's funds, but underpinned by estimates of the opportunity cost of capital in developing countries (World Bank 1996, Technical Appendix, paragraph 49). This was, however, omitted from the published version (Belli *et al.*, 2001). Various arguments and rates are discussed in Penman (1999).

16 Zerbe and Dively (1994, Sections 13.4.1, 13.4.4) arrive at a range of social time preference rates of 2.5–5.5%, for the USA, by adjusting specific kinds of market interest rates.

17 Ethical considerations are particularly dealt with by Hanley and Spash (1993, Section 8.3) and Price (1993, Ch. 12).

18 Price (1993, p. 200), Pearce (1994). In effect Kula's special discounting scheme discounts in the normal way for early decades of a project's life, but the effective discounting factors eventually level off, not decreasing further.

19 H.M. Treasury (2003), Hanley and Spash (1993, Section 8.4, p. 144). See also Pearce and Turner (1990, Ch. 14), who explain why using low discount rates does not necessarily favour environmental concerns. Roberts (1996) argues for using a relatively high discount rate, but assuming increasing value of some environmental goods in the long term (escalation), which has the arithmetic effect of applying a lower discount rate to those effects. Pearce (1993, Ch. 3) argues for a rate in the range 2–5%. Abelson (1996, Section 2.6) describes a way of applying the opportunity cost of capital to the capital costs and the social time preference rate to other elements, linking this to the earlier 'shadow price of capital' approach. Pearce (1994) explains how the special low discount rate for forestry should not be applied when non-market benefits are fully included in the analysis. Lind (1999), among others, considers the issue of future generations being richer than current ones. Zerbe (2004) says people are willing to pay for moral values about intergenerational equity, but that it may be better to apply such moral values directly and explicitly rather

than just using very low discount rates. The revised British government guideline (H.M. Treasury, 2003) sets a rate of 3.5%, explicitly derived as a social time preference rate, for all sectors, but requires lower rates for long-term analyses, down to only 1.0% for years beyond 300.

20 Hanley and Spash (1993, Section 8.2.1, p. 131), Cline (1993). For the USA, Cline suggests a social time preference rate of 1–5% and an opportunity cost of capital of 8%; using a weighting method (ABFK), he arrives at an appropriate discount rate for all sectors of 2%, but says it would be higher in a poor country with rapid growth (his social time preference rate is derived entirely from growth and marginal utility of consumption, with zero pure time preference).

21 Modern CBA derives largely from decision guides developed for the US water resources sector in the 1930s to 1950s.

22 Published in the UK by Beech Tree Publishing.

F

Multicriterion decision analysis

F.1 Introduction

Most decisions need to be made by reference to several criteria or objectives. I choose a new bicycle or car on grounds of cost, comfort, expected long life, vulnerability to theft and appearance. Planners and designers might choose a dam site on grounds of cost, size of available storage, annual inflow, location relative to water demand centres, environmental impact (including visual merit or demerit, affect on flora and fauna, etc.) and social impact (number of homes, farms or factories needing to be relocated).

In general, a decision-making problem or situation will usually boil down to:

- a number of alternative courses of action (variously called options, projects, schemes, interventions, packages, plans, or variants according to context)
- a number of objectives or criteria
- one or more persons or groups of people to whom the criteria matter to some degree.

Decision-making methods can be complex or simple. If we try to take into account every possible nuance associated with a decision, we may not get round to making it and implementing it until the relevant opportunity has passed, so we have to simplify it to some degree. The extreme simplification is to limit oneself to one criterion: I might buy the fastest car that I can afford (regardless of cost or comfort), or build the biggest dam that the available sites offer (regardless of environmental impact, etc.). Financial CBA approaches this degree of simplification, by basing its decision indicator on money alone. Economic CBA, if and when it takes in wider criteria by means of shadow pricing, is a little more subtle. There are many other decision-guiding methods and approaches, or decision support systems, which explicitly cover more than one objective or criterion. They are referred to under several names, such as *multicriterion, multicriteria* and *multi-attribute*. A vast literature has built up, but many decision-makers and analysts are scarcely aware of it. The purpose of this appendix is to introduce the reader to multicriterion approaches, as a relevant part of the context within which CBA operates, to describe one simple but powerful method, and to point to sources of further information. Readers needing further guidance could use Belton and Stewart (2002), or the freely available British government guideline (DCLG, 2009). The latter emphasises that multicriterion analysis is a complement to CBA, not a substitute, and includes examples.

If decision-makers do not understand the analysis by which their decisions are guided, they are unlikely to 'own' the decisions, and the decisions are unlikely to be effectively implemented. Similarly, if other affected people or groups are not clearly informed of the reasoning behind a proposal, they cannot give it effective consent or act to improve it. Once a decision has been made, it usually has to be explained and defended to a wider audience, so that a process of communication and persuasion must follow the decision-making. For all these reasons, a decision aid must often serve as an instrument not only of analysis but also of communication and persuasion. Therefore, one of its most important characteristics must be *transparency*. If an analyst collects information, processes and analyses it, she must then present a clearly argued case to the decision-makers and other people who are to be consulted. All these people must understand the analysis well enough to influence it, asking the analyst to modify it until they are comfortable with it. A non-transparent analysis can be used by an analyst to usurp the decision-makers' role, by hiding undeclared priorities and biases in a *black box*, which is either not described at all or described in impenetrable technical jargon, but such an analyst is unprofessional and his work is not conducive to good decisions.

Section F.2 introduces some basic terms and concepts. Section F.3 describes some non-numerical methods that are potentially useful and seldom mentioned in the multicriterion literature. Among the numerical methods, Section F.4 describes a simple multicriterion analysis method, and Section F.5 sets out the general framework within which that method is a special case, and briefly describes some other approaches. Section F.6 considers uncertainty, while pointers to further reading and sources are given in section F.7, as well as in the Notes.

F.2 Basic concepts and techniques

F.2.1 Terminology

People have approached multicriterion analysis from many different backgrounds and situations, often unaware of what others had been doing and thinking, so there is no single agreed set of terms for the various concepts. In this appendix the alternative courses of action will usually be called *options*; in engineering or project design contexts they are often called 'alternatives' or 'variants', and in planning or investment-decision contexts they are often called 'schemes', 'projects' or 'plans'. With regard to the second of the three sets listed above, they are here referred to as *criteria*, but related terms (not synonyms) are attributes and objectives. The third set are here called *interest groups*.

Other concepts also have several names, some of which are introduced below; what I call a *scoring rule* is, for example, also called a *value function* or a *transformation function*.

F.2.2 Scoring or evaluating

Before any arithmetic can be performed in any of the numerical methods, the characteristics of the different options need to be expressed in numbers. Typically, the merit of one option under one criterion is given a score between 0 (the undesirable end of the scale) and 100 (the desirable end). Some common kinds of scoring rule are described below, the first three types being illustrated in Figure F1.

(a) *Linear functions* of some physical or monetary parameter, such as cost, number of patients treated, volume of water provided or passenger-miles. A certain range of parameter values is chosen (it can be the range covered by the options currently available for analysis, or some wider range arbitrarily designed to make room for more extreme options that might show up later). The range limit on the undesirable end is usually assigned the score zero, the other limit is assigned the score 100, and all parameter values in between get a score by linear interpolation. (People who use formulae have been known to get their scoring upside down, assigning the most desirable score to the highest cost: common sense must override neat mathematics.) One disadvantage is that if an option shows up late in the analysis process and has a parameter value outside the chosen range, the range will have to be redefined and all values recalculated to avoid going outside the 0 to 100 score scale.

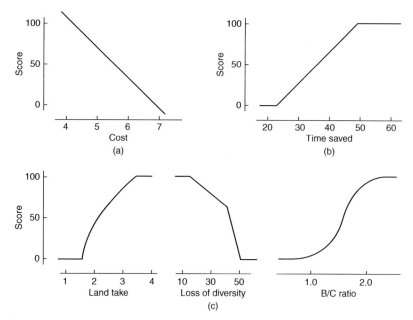

Figure F1 Some types of scoring function: (a) linear; (b) truncated linear; (c) truncated non-linear

(b) *Truncated linear functions* are like the plain ones but stipulate that any parameter value beyond the end of the chosen range will have the value 0 or 100, as appropriate. This not only avoids the above problem with late entrants but also enables the analyst to choose a range that is really of interest, whatever the range displayed by the set of options that are up for analysis. For example, if a CBA indicator were the parameter for the criterion in question, the analyst could decide that any option with an EIRR of more than 20% would be assigned the value 100, implying that if an option is that good economically she does not care if its EIRR is 21% or 31% or even 41%. Functions can be truncated at one end or both ends.

(c) *Non-linear functions* are similar to linear ones except that the analyst or user decides that one cannot assume that equal numerical increments of the raw parameter correspond to equal increments of desirability, which is what the score is trying to measure. For example, some of a set of proposed irrigation schemes would serve subsistence farmers with average farm sizes as small as 2 ha, while other schemes would benefit mainly farmers with farms larger than 20 ha: the decision-makers wanted a farm-size criterion in their analysis so as to bias the choice of schemes in favour of those serving small farms. A linear scoring rule to convert average farm size into a desirability score would make the difference between 20 ha and 22 ha just as important as the difference between 2 ha and 4 ha. If this seems inappropriate to the objective, one might prefer to say that a doubling of farm size would have the same significance, and hence the same score differential, anywhere in the range: a pair of schemes with farm sizes of 2 ha and 4 ha would then have the same score difference as another pair with farm sizes of 20 ha and 40 ha. Arithmetically this sort of 'equal ratio, equal score' difference rule is a logarithmic one, and it is almost as easy to compute as a linear rule. One can also use multiple linear functions (graphically, a kinked function of straight lines like the middle one in Figure F1(c)), or S-shaped curves if the importance of equal parameter increments is highest in the middle of the range and tails off towards both ends. Non-linear scoring rules can be truncated or not, and can use almost any shape provided that any parameter value gives one and only one score.

(d) *Additive* or *weighted-total scoring rules* are helpful when a criterion or attribute is itself a compound of several elements. When comparing dam sites in 1992, we needed to express the degree to which their reservoirs would displace buildings, but there were different sorts of building which were not equally important. First we calculated for each dam site the number of equivalent homes displaced, where a real home counted as one unit, a farmstead with outbuildings as two units, and a holiday home (not occupied all the year round) as half a unit. The total number of units for each dam site was then converted to a score by means of a simple linear rule.

(e) *Verbal scoring rules* are used when the parameter or attribute cannot be numerically measured at all. For instance, in choosing between many different water-transfer schemes we used, as one criterion among several, their susceptibility to sabotage and vandalism. The extremes of the scoring rule were defined as:

> *if a scheme shows very little susceptibility, because all elements are easy to guard and difficult for unauthorised people to get at, the score is 90; a highly susceptible scheme, like one whose pipelines, intakes, pumps and their power supplies are widely dispersed and open to public access, would score 10.*

In between, four examples were given to guide the assignment of scores, such as:

> *if the system is mostly deep underground but there are two isolated pump stations, the score is 40.*

The six verbal descriptions were shown on a diagram against the 0 to 100 scoring scale, and scores were assigned to 60 schemes quite quickly. The subsequent report showed the rule diagram and the scheme scores, so that any reader could check that the scoring rule gave sensible results.

The scoring rules with algebraic formulae and graphical equivalents tend to look more scientific and objective than the verbal ones, but in fact all these scoring rules are subjective. The choice of linear or some other shape, and the choice of range limits, is as subjective as the definition of the weights in an additive rule or the words and their scores in a verbal description. Any kind of scoring rule can be set by an analyst working alone, or set by a committee, or determined by some wider consultation process, but none is objective like a measured temperature or length.[1]

F.2.3   Weighted averages and totals

The common arithmetical device of weights applied to different categories is often useful in multicriterion analysis. The concept of equivalent homes in the previous section is an example of a weighted total. A *weighted average* is derived by multiplying contributions from different categories by some sort of importance weights before computing an average. Suppose three trade-union leaders want to set the thermostat in their conference room but insist on their relative importance. The first represents 10 000 plumbers and wants the room at 20°C, the second represents 20 000 electricians and prefers 24°C, while the third represents a mere 500 airline pilots and wants a cool 16°C. If they are serious about weighting all decisions by union membership, they will set the thermostat at 22.6°C ($20 \times 10\,000 + 24 \times 20\,000 + 16 \times 500$, all divided by the total membership of 30 500). The pilots' leader has only 1.6% weight in this decision (and any other decision taken in the meeting, presumably) and must either accept his lot and remove his pullover, or else challenge the membership-weighting rule.

F.2.4   Dominance

Suppose 12 options are being compared by the use of five criteria, to find one preferred option. If option B is better than option E on every single criterion taken alone, then we say that option B dominates option E. The useful consequence is that option E can be dropped from the analysis

217

before we even begin to consider the other 10 options or the relative importance of the criteria. If we do remove option E and any other dominated options from the list, the remaining options form the non-dominated set. (However, in reporting an analysis we may choose to keep the dominated sets in our lists and tables so as to inform people that they were considered and show why they were not favoured.)

F.2.5   Hierarchies

Objectives are often hierarchical, and some people use different words such as 'aims' or 'goals' at distinct levels of the hierarchy. A useful device for developing a set of criteria and their importance weights is a *value tree*, which describes a hierarchy of objectives grouped at two or more levels. Figure F2 shows an example of a value tree that might be used for guiding decisions about irrigation projects that compete for limited funds in a country with security problems.

In the example described above, the criterion of susceptibility to sabotage and vandalism was one of four criteria in a group defining reliability, which was used alongside five other groups in an analysis that used 16 criteria.[2] At the upper level of that hierarchy there were six objectives: cost, reliability, ease of implementation, minimisation of land take, preservation of fisheries, and enhancement of the environment (other than fisheries). The reliability objective had four sub-objectives and so did the environmental one, while the others had two each. Organising the objectives into a hierarchy corresponds to organising the criteria into groups. Some analyses use as many as four levels.[3] This grouping or subdivision of objectives is very helpful when there are many objectives, because the decision-maker or affected person can consider the importance of a group relative to other groups, and then as a separate exercise think about the importance of the group's constituent elements relative to each other.[4]

F.2.6   Screening, ranking and thresholds

Screening and ranking are two basic and complementary processes used to reduce a long list of conceivable courses of action or options to one preferred option, or to a preferred set of options that matches a budget or any other constraint. *Screening*, by analogy with a sieve that lets some items through and holds others back, is the simpler: an option is either accepted or rejected at a screening phase. It might be rejected by being shown to be impracticable, or totally unacceptable

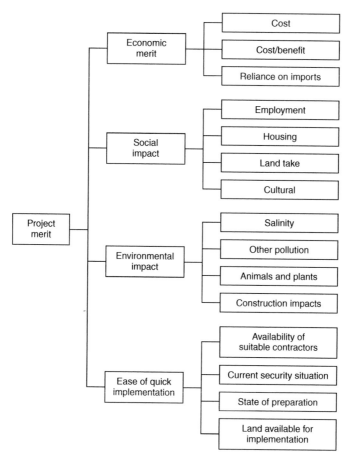

Figure F2 An example of a value tree

under just one criterion (also called a veto), or just by falling below some more or less arbitrary threshold based on one or several criteria. A threshold is an essentially non-linear and discontinuous concept, but can be mixed with smooth relationships, as, for example, in truncated scoring rules.

Ranking is more subtle than screening in that options are not merely sorted into two classes, the accepted and the rejected, but are placed in order of merit, by whatever measure of merit or preference is being used. Suppose three options A, B and C achieve the values 8, 5 and 9, respectively, on some merit or preference scale. One can say that the respective ranks are 2, 3, 1, or that the rank order is C, A, B. Either way one is omitting some information, namely, the fact that the preference for C over A is by only one unit of preference, while A is preferred to B by a margin of three units. The use or reporting of ranks only, when preference margins are known or estimated, thus involves a loss of information. Conversely, there may be situations of limited information when statements can be made about rank order, even when nothing can be said about margins.

F.3   Non-numerical methods

Before describing the numerical ways of combining several criteria to guide a decision, it is as well to put them in context by mentioning some others. Two that are useful in practice are:

(a) *Pairwise comparisons*
This simple approach involves taking the available options in pairs, and, by examining the main differences of nature and impact, determining which of each pair is preferable. Once all the pairs have been analysed, one can check for incompatibilities (e.g. A being preferred to B and B to C, but C preferred to A) and adjust accordingly to reach a useful, communicable and defensible conclusion. This approach is sometimes attractive when there are few options, and allows the use of different criteria for different pairs, which gives it flexibility. The analysis is, however, less structured than with other methods, and is sometimes difficult to defend against an objector who sees it as biased. It cannot deal with trade-offs, when different criteria favour different options.

(b) *Systematic verbal comparison*
This method is good when the numbers of options and of criteria are moderate and the options are very diverse, so that some verbal explanation is needed to explain why a particular option is assigned a particular description under a particular criterion. Usually the presentation is by means of a family of similar tables, one per option or one per criterion. I used this method to guide a choice of disposal site for slightly toxic sludge dredged from urban waterways. Each of nine sorts of site, as diverse as deep sea dumping, disused brick pits and islands in the waterways, was examined in terms of a standard set of eight criteria such as access, public health impact and historical/cultural impact. The report contained a table for each type of site, in which the impact under each criterion was described in up to 60 words and its significance then graded as serious, undesirable, moderate, neutral/negligible or favourable. Finally, the analysis of all nine types was summarised in one table, and a recommendation was distilled out verbally, without any numbers.

Non-numerical methods are akin to the numerical ones discussed in the rest of this appendix; indeed it would be possible, in principle, to adapt the disposal-site analysis just described and make it a numerical analysis.

F.4 A simple numerical method

As an introduction to the numerical methods as a whole, this section describes a simple one, which is appropriate for many tasks. It is called the additive model within multi-attribute utility theory, or the *weighted-average method*. Because of its simplicity, the principle is easy to explain to decision-makers and the wider public. In terms of the sets identified in Section F.1 it deals with the first two only, in that the selection problem is structured as a list of courses of action (options) and a list of criteria. This method does not consider divergent points of view as separate sets, although it can incorporate them in the handling of the one set of criteria.

For instance, suppose that a supply of potable water is to be made available to a town. There are two schemes or options:

- construct a well-field in a distant aquifer (which has good water but limited capacity), and pump the water to the town
- abstract water from a large nearby river and treat it.

The criteria are cost, reliability and environmental impact. The first step is to define a scale for each criterion so that the characteristics of each option can be quantified. In this example the scores all run from 0 (the undesirable end) to 100, as described above. The scoring rule can be numerically based or subjective.

Suppose the range of conceivable costs for the water supply (as present values of capital and recurrent costs) is £5 million to £9 million: these might be assigned the score values 100 and 0, respectively, and the scoring might be defined as linear. Then the groundwater scheme, with a present value cost of £8 million, would receive the unfavourable score of 25, and the river scheme costing only £6 million would be scored at 75.

Reliability cannot be scored by calculation, so a verbal type of scoring rule is used. Suppose that reasoning like that in Section F.2.2(e) is used. The groundwater scheme scores only 30 points because of the long pipe-line, several pumping stations with long power lines and the risk of water shortage in dry spells. The river scheme scores 65 points because its mechanical and electrical equipment is all close together and near the town, while the river always has enough water.

Environmental impact also requires subjective judgement. The distant aquifer is not related to drying up of chalk streams, and with the pipeline buried the only adverse impact is the minor visual intrusion of the power lines. The river scheme, however, would involve an intake in a particularly beautiful area, beloved of fishermen and frequented by bird watchers. It would also significantly reduce the dry-season flow downstream. For the purposes of this example, we suppose that some scoring rule, perhaps additive as described above, assigns scores of 80 and 35 to the two schemes.

The other element needed in order to complete the weighted-average analysis is the relative importance of the criteria, quantified as a set of weights. In principle, they are like the union membership figures in the earlier example, although they only have meaning relative to each other, and it is convenient to make them add up to unity or 100%. The two setting-up processes, weighting and scoring, are not independent of each other, and the whole analysis needs to be developed in an iterative manner, trying out early versions with real alternatives, before it gives a good representation of the decision-makers' preferences and value judgements. In this simple example, the analyst, after talking to the decision-makers and thinking about other affected people and their known or surmised preferences, tentatively sets the importance weights at 40%, 25% and 35% for cost, reliability and environmental impact, respectively.

The analysis consists of setting out the options and the criteria in a table, entering the scores and criterion importance weights, and calculating the multicriterion index for each option, in this case for each water-supply scheme. This combined index for each option is just the weighted average of its scores under the various criteria, using the criterion importance weights (Table F1). Like the scoring rules of which it is a weighted average, the index runs, in principle, from 0 (undesirable) to 100 (desirable). Screening or ranking of possible options is done by comparing the multicriterion index of one option with that of another. If desired, the tabulated options can easily be

Table F1 Example of a simple multicriterion analysis

Criterion:	Unit cost	Reliability	Environmental impact	Multicriterion index
Importance weight	40	25	35	
Groundwater scheme	25	30	80	45.5
River scheme	75	65	35	58.5

sorted into a ranked list, in order of merit according to the criteria and weights that are being used.

In this two-scheme example the analysis shows a clear preference for the river scheme, with a multicriterion index value of 58.5 against a value of 45.5 for the groundwater scheme. In reporting or explaining the analysis the reasoning can be described verbally by reference to the table:

> *the groundwater scheme's environmental advantages are outweighed by the river scheme's lower cost and better reliability.*

This is termed a *trade-off*, where a disadvantage under one criterion is accepted in view of an advantage under another. As in this case neither option dominates the other, some compromise or trade-off is needed, and this way of presenting the analysis makes it explicit and clear.

A transparent analysis of this kind is conducive to consultation. Suppose that the analysts now go to the decision-makers with this table as a draft or suggestion. The decision-makers in the first instance are the water resources authorities, and the low-cost and high-reliability criteria are naturally in line with their objectives. They are, however, responsible citizens and realistic planners, so they also want to apply the wider environmental criterion. They therefore ask the analysts to give more weight to the environmental impacts: after discussion the weights are changed to 25, 25, 50 (the same weight as before for relia- bility, but considerably less for cost and more for environment). In a matter of minutes the analysts can report that the multicriterion index values are now 53.8 for the groundwater scheme and 52.5 for the river scheme. The higher weight now given to environmental issues is just enough to tip the balance in favour of the groundwater scheme. Other sets of weights can easily be tried, as sensitivity tests or as part of a discussion to elicit weights from decision-makers.

In this example of decision-maker participation in the analysis, only the criterion weights were changed. It is also possible, and very right and proper, that the scoring rules may be questioned. The spread of index values under each criterion needs to be reasonable: if one criterion has the values for all options bunched up in a small part of the range, say between 50 and 70, the criterion's importance weight will in effect be lowered. Ideally, equal increments of index should represent equal increments of desirability or undesirability, over the whole range for any one criterion. There are statistical techniques for eliciting weights and scoring functions from decision-makers, but it is often sufficient to follow this simple approach: the analyst suggests a set of rules and criteria weights, the decision-makers discuss and amend it, and if necessary the

223

analysis can be reworked several times until everyone feels comfortable with it. It is generally best to begin by agreeing which is the most important criterion and giving it an initial weight of 1, and then to discuss the weight to be given to the second most important criterion relative to that. The weight is not just people's intuitive idea of the importance of the criterion, but must be assessed by asking what interval on one criterion would be seen as a fair trade-off against a standard interval (such as the full range 0 to 100) on another criterion. Belton and Stewart (2002, p. 115 and Sect. 5.4) describe this 'swing weight' method. This is why the scoring rules should be developed and agreed before assessing the weights. Usually the weight of the most important criterion should not be more than five times that of the least important, as a greater ratio would give the least-weighted criterion very little influence on the overall merit index; an exception can be made when criteria with small weights have vetoes. The transparent and iterative way of arriving at the elements of the analysis helps to build up the decision-makers' confidence in the method, which can otherwise be eroded by the application of an obscure statistical technique. In any case, the activity of setting up scoring rules and then filling in the scores is itself an illuminating one, making the analyst and decision-makers consider consciously a large number of aspects of each option. It is impossible to complete this process without becoming well aware of the inevitably subjective nature of the analysis, which is a good antidote to the delusions of precision or objectivity that many analysts are prey to.[5]

This imaginary example serves to demonstrate the simplicity of the weighted-average method. It can handle complex analyses without losing its essential transparency. One application, for selection of a water source from 60 alternatives, involved 16 criteria in a first stage that reduced the list of alternatives to 22, then 23 criteria in a second stage of analysis.[6] The 23 criteria were grouped at three levels and some were additive functions of subcategories at a fourth level. Coloured diagrams were used to express the results, and 16 sensitivity tests were carried out on the weights (using a large but simple spreadsheet with automatic ranking); the reporting of the second stage alone covered more than 50 pages in the report. Trade-offs were demonstrated by using the grouping of criteria (hierarchy of objectives) to tabulate group contributions, and were explicitly discussed with words such as 'relative', 'outweighed' and 'despite'. Decision-makers with no prior experience of formal multicriterion analysis welcomed the method and were contributing to the setting of weights within minutes of its first presentation. A complex problem was thus tackled with a simple tool, the primary merit of which was transparency.

It is advisable, when setting the scores for the various options, to keep a record of the reasoning behind the decisions, so that people can come back later and be reminded why a certain score was set. Simple standard forms are suitable. This is helpful when setting scores for a new set of options at a later date. A set of such record sheets contributes to the definition of the scoring rule.

This sort of analysis can use criteria or attributes that are not totally independent of each other, even ones that are obviously correlated because of a causal link between the attributes, or a common cause underlying two of them. All that is needed is to ensure that the people who set the criterion importance weights are aware of the correlations, and the grouping of criteria helps with this.

F.5   Other methods

The general framework of multicriterion analysis was set out in Section F.1. The simple method explained above is a special case, and there are other approaches. Some are described briefly here.[7]

(a) *Satisficing, or lexicographic ordering methods*
In these methods the criteria are taken one at a time and used to screen out options. All options are examined under the first criterion and some are rejected, then the remainder are examined under a second criterion, and so on. The method obviously only works if the criteria involve rejection/acceptance as well as mere relative merit. The basic method does not take trade-offs into account, and may not produce a single recommended option. The method can be used in a preliminary screening stage, to eliminate unacceptable options before using another method to analyse the rest with reference to trade-offs. Implicitly this is often done without any reference to a method, in that the analyst will identify unacceptable options at the option formulation stage and not take them forward to the multicriterion analysis at all. One can also incorporate a veto in almost any kind of multicriterion analysis, as a feature of a scoring rule.[8] This group of methods is more a description of how people sometimes make decisions with limited information than a recommended method, although it can be useful for initial screening of a large number of options.

(b) *Distance-from-ideal approaches*
Given a set of options, each numerically evaluated under a number of criteria, one can define a hypothetical ideal option or benchmark that has, under each criterion, the best score seen for that criterion in the

225

whole set of real options. One can then compute how far each real option is from the ideal option, and use this distance measure as an index of multicriterion merit. With two criteria this can be shown on an ordinary two-axis graph, with three it can be imagined as a distance in three-dimensional space, and with more than three criteria it can (with increasing strain on the less mathematical minds) be imagined as a distance in multidimensional space. The relative importance of the criteria is not explicitly handled. The definition of the distance measure includes an arbitrary constant whose significance is difficult to grasp or to explain. The method lacks transparency and seems to offer no significant advantages in return.[9]

(c) Rank-arithmetic approaches

This is a set of methods that pay particular attention to the differing preferences of distinct interest groups, the third set in the general description of the problem in Section F.1. Some simple methods begin by ranking the options separately according to each group's preferences, and then combining the results by simple arithmetic on the rank numbers. This implicitly assumes that preference is uniformly distributed by rank, with equal margins of preference, which is obviously wrong: I might marginally prefer marmalade to jam, but prefer both to peanut butter by a large margin, yet this method would give equal weight to the marginal preference and the heavy one. Differing importance of interest groups is also not covered. There are other technical problems, and these rank-arithmetic methods are generally not useful.[10]

(d) Additive approaches

This is the group of which the weighted-average method, described earlier, is a member. Besides the important merit of transparency emphasised in Section F.4, they are good for dealing explicitly with *compromises* or *trade-offs*. The existence of several interest groups with different sets of criteria and/or criterion weights can be dealt with in a number of ways. One is to repeat the options–criteria analysis for each group's weighting set; by applying a weight to each interest group one can then still arrive at a single-number index for each option. Another way is to group the criteria according to the main concerns of the distinct interest groups, and then vary the weights within the options–criteria analysis. Either way, sensitivity tests are a powerful tool in the search for consensus or agreement, or for explanation and persuasion.

Although the results can be discussed in terms of rank, the margins by which options outrank each other are always known as well, so these methods do not throw away information on the relative strengths of preferences. Straightforward additive methods such as the weighted-average one are essentially linear, but scoring rules can include thresholds, sudden slope changes, and almost any kind of non-linearity or discontinuity that may be needed in order to represent a real situation, without restrictions from the analysis method.

There are a number of methods and variants within this class, some associated with particular suites of computer software. In many cases the special features consist in the treatment of uncertainty. The common feature is that a single indicator is produced, and this itself is sometimes criticised. (The first time I used multicriterion analysis in the form of a table of scores against options and criteria, I initially wanted to stop at that and not compute the multicriterion index, on the grounds that the bare table of scores would force the decision-makers (who were inexperienced officials just emerging from decades of rigid central planning) to think about each criterion separately rather than uncritically accepting the magic number computed by supposedly clever consultants. After the first round of discussions I was persuaded to relent and compute a multicriterion index after all.) It is important to insist on transparency at every stage and to emphasise that every element, from choice of criteria, to design of scoring rules, to weights and the design of sensitivity tests, is subjective and demands to be examined and questioned by everyone interested in the outcome. Sensitivity tests are always desirable.[11]

(e) *Concordance and outranking methods*
This is a family of methods, mostly with associated computer software, originating from the work of Bernard Roy in France since the 1960s. In the basic method, starting from the set of scores for several options under several criteria, two index values are computed for each pair of options: the concordance index takes account of all criteria under which the first option is preferred to the second, and of the weights of those criteria, while the discordance index relates to only one criterion, namely that under which one option is most strongly inferior to the other. Thus the concordance index ignores information on strengths of preference under particular criteria, while the discordance index ignores relative criterion weights. The concordance and discordance values for all options are then used to determine the best option, in a complicated iterative procedure involving variable thresholds. The methods can be seen as analysing degrees of dominance.

Many variants have been developed from the basic method, some of which use the concepts of distance from one or more benchmarks, as in the distance-from-ideal methods mentioned above. Some methods produce a full ranking of options with an indication of relative margins of preference, like that produced by the weighted-average method, but some do not. These methods are generally hard to understand and therefore to explain to decision-makers and affected people; they lack transparency. Nevertheless, various merits are claimed.[12]

(f) *The analytic hierarchy process*
This method breaks down the decision problem into parts in a hierarchical manner and makes pairwise comparisons. it then builds them up into indications of overall preference by complex mathematical techniques.[13]

(g) *Methods based on CBA*
Various attempts have been made to extend CBA to cover the multi-criterion problem. The *planned balance sheet* concentrates on the distinct impact of options on different interest groups, and is in that respect similar to the corresponding emphasis, notably by the World Bank, in normal single-criterion CBA. The *goal achievement matrix* concentrates on the relationship between objectives and interest groups for each option, but using the restricted indicators of CBA instead of the more general scoring concepts described in this appendix. Such methods have limited application.[14]

The choice of multicriterion analysis method depends on the situation, especially the nature of the criteria, the numbers of options, of criteria (objectives), and of distinct interest groups, and the amount and quality of information available. Many decision-makers and interest-group representatives will be confused or antagonised by mathematical complexity, and the simplicity of the simple weighted-average method will often be a decisive factor. The simple additive method can handle large numbers of options (even in hundreds) and of criteria (if grouping of criteria is used), but is not particularly well adapted to dealing formally with many interest groups. On the other hand, its transparency makes it amenable to wide consultation, which enables many people to influence the outcome without needing to be formed into labelled groups, and this is itself an advantage. Concordance and satisficing methods are better for screening or short-listing than for final selection of one preferred option.[15] They can operate when information is scarce, and can arrive at a result without analysing trade-offs, thus without needing criterion importance weights, which may be useful when interest groups cannot agree on those. However, any method that manages with less information than another method usually does so by making assumptions, whether implicit in the method or explicit as values of constants or thresholds, and this may negate the apparent advantage. The analytic hierarchy process may be preferred where setting the criterion importance weights is particularly difficult, but it lacks transparency.

The recommendation for readers of this book who need to conduct a multicriterion analysis is either to use a non-numerical method such as those described in Section F.3, or to use the simple weighted-average method, unless there is a pressing reason to do otherwise, such as the need to conform to a client's requirement for comparability with another analysis. Most of the situations that the other methods were developed to deal with can also be handled using the weighted-average method. If information is scarce, for instance on criterion weights for some interest groups, it may be best to start by guessing the numbers and show a transparent analysis to the affected people, inviting them to suggest improvements.

F.6   Uncertainty and incomplete information

Decision-makers never have complete and precise information about the available options or their merits under various criteria, nor about the relative importance of the criteria. As well as partial information, there are usually divergences between the preferences or value judgements of the different decision-makers or stakeholders. Any kind of multicriterion analysis, to be useful for real decisions, must deal with these uncertainties and divergences. Some experts have developed ways of using fuzzy set theory and rough set theory (lower and upper bounds) to deal with uncertainty, while the satisficing and outranking approaches have their own internal ways of dealing with it.[16]

A common and useful way of dealing with uncertainty is to do sensitivity tests in discussion with stakeholders. After a model or computation has been set up with the most likely value of each important input parameter, it can be re-run with different values to see what difference each change makes. This can guide the iterative improvement of the model, and if decision-makers are involved at successive stages it can help them to formulate their preferences clearly and to reach a consensus view.

F.7   Further reading and information sources

Some references to publications relevant to specific topics have been given in the Notes, the full references being in Appendix K; each of these contains further references. Readers wishing to find out more about multicriterion analysis in general, or needing detailed advice, may find the following books useful:

- DCLG (2009): the British government's thorough manual for multi-criterion analysis, especially regarding environmental aspects of public projects and programmes.
- Belton and Stewart (2002): a thorough and useful book by two leading academics in the field.
- Zeleny (1982): for a broad discussion, especially Chapter 14: Decision Making is a Very Human Business, covering 'Information and Confidence' and 'A Mind Snapping Shut'.
- Watson and Buede (1987): for a thorough discussion of the psychology of decision-making and a wide range of analytical approaches.
- Massam (1988): for a broad-ranging summary in 80 pages.
- French (1989): for a general description, several case studies and many further references.
- Goodwin and Wright (1991): for clear expositions, particularly of scoring, additive methods and uncertainty (linking with Appendix A in this book).
- Vincke (1992): for a general review with emphasis on concordance methods.

More recent publications can be traced by referring to papers and book reviews in the *Journal of Multi-Criteria Decision Analysis* (Wiley; formerly called *Journal of Multi-Criteria Analysis*).

Notes

1 The iterative process of eliciting a subjective non-linear scoring curve (value function) from a decision-maker is described in Belton and Stewart (2002) and in Goodwin and Wright (1991, Ch. 2).
2 This example, and some others, is described more fully in Snell (1994).
3 An example with four levels is described in Stansbury *et al.* (1991), with the terminology aggregation of indicators, and one with three levels in Snell (1994), called grouping of criteria. Belton (1989) calls the same technique a hierarchy of objectives. Belton and Stewart (2002 especially pp. 66, 80, 139) discuss hierarchies and groups. DCLG (2009, Sections 5.4.3, 6.2.6) covers grouping and value trees.
4 See Belton and Stewart (2002) and Goodwin and Wright (1991, Ch. 2).
5 The weighted-average or additive method, along with ways of eliciting preferences and priorities from decision-makers, is discussed in Belton and Stewart (2002) and Goodwin and Wright (1991).

6 This example is described in Snell (1994), from which the above imaginary example is also adapted.

7 The order and terminology generally follows that of Massam (1988).

8 In one screening and prioritising analysis the scale was defined so that the values 1 to 99 were the scale of relative merit, while zero represented a veto and 100 a compulsion. A plan with a zero score under any criterion was rejected, however good it was under other criteria, and one with a score of 100 under one criterion was accepted, regardless of the other criteria. This amounts to a decision to include in the analysis some options which could have been dealt with outside the analysis; it is purely a presentational device.

9 Massam (1988) calls this class 'graphical approaches'. It includes a method named *TOPSIS*. A variant compares distances from a *status quo* set of scores rather than from an ideal. A method of this type, called *composite programming* algorithm, is described by Stansbury et al. (1991).

10 These methods are described in Massam (1988, Section 3.4) under the name *consensus maximisation*.

11 Massam (1988, Section 3.5) describes the additive methods, and most of them are often classed as variants of *multi-attribute utility theory* (MAUT); a good description is given by Goodwin and Wright (1991, Ch. 2). Rogers and Bruen (1995) call this class 'index methods'.

12 Massam (1988, Section 3.6) gives a summary, and there are many books and papers on these methods, especially by Vincke (1992) and by two others named Nijkamp and Roy. The named methods include ELECTRE I, ELECTRE II, ELECTRE III, ELECTRE IV, ELECTRA, SELECTRA, PROMETHEE and REGIME.

13 The techniques involve eigenvalues at several levels, and are described briefly by Rogers and Bruen (1995), who also give references to fuller descriptions.

14 These are also described, with further references, by Rogers and Bruen (1995).

15 Rogers and Bruen (1995, Paragraph 8).

16 See DCLG (2009, Sections 4.8–4.11).

G

The effects method

The effects method is a way of analysing and appraising development projects, which is in some ways an alternative to CBA as described in this book, and in some ways a complement to it. The method was developed in the 1960s for use in French-speaking African countries, and that group of countries has remained almost the only place it is used.[1] This appendix describes it only briefly, for two reasons: firstly because it is little used, and secondly because it is more difficult for non-economists to understand than is normal CBA. Readers needing more information will find it in the works cited in the Notes; full references are given in Appendix K.

The effects method analyses a project by calculating its financial effect on the economy in terms of value-added, which in turn measures the effect on gross domestic or national product; it is thus associated with national accounting.[2] The gain to the economy as a result of the project's implementation is evaluated, for every year of an analysis period, by quantifying three elements of incremental value-added:

- direct value-added generated by the project, including effects on people receiving its products (downstream effects)
- indirect value-added generated through the project's effect on its suppliers (upstream or backward linkages)
- incremental value-added due to differences between the with-project and without-project situations.

Usually the resulting annual figures for total incremental value-added due to the project are set against the investment costs and discounted in the same way as for CBA, to produce an IRR. Sometimes a more subtle value-added approach is used on the cost side also. Evaluating the upstream effects requires some combination of:

- analysing the operating accounts of suppliers, and perhaps their suppliers (which may be difficult to obtain access to), and

- applying known linkage coefficients for the economy (available in some countries in the form of an input–output table, but these are seldom accurate and up to date).

The main characteristics of the method are that:

- it uses financial prices but, if done thoroughly, traces all effects back to the national economy, particularly border effects (imports and exports and foreign exchange balance)
- it separately identifies the project's impacts on different parties, i.e. distributional aspects.

The absence of implicit shadow or economic pricing makes the method appear to be a financial one, although it can be argued that the tracing back of effects to the border, i.e. to national opportunity costs, makes it an economic analysis in the efficiency-measuring sense. Economic theorists can thus demonstrate that the effects method is formally equivalent to economic CBA.[3] It remains, however, a very distinct procedure which can normally only be carried out by economists who have been specially trained for the purpose. Some regard it as a complement to economic CBA which should be done alongside that CBA. One suggestion is that the final decision guide should be a multicriterion analysis using economic CBA to measure efficiency, the effects method to measure equity (distribution of costs and benefits), and financial analysis of different parties to measure sustainability and cost recovery.[4] Many decision-makers and financing agencies,[5] however, now advocate using the party-specific financial analyses to measure equity, which leaves no need to do a time-consuming effects method analysis alongside the others.

Some decision-makers, generally only in Europe and French-speaking Africa, require their projects to be analysed by means of the effects method, and an analyst will not normally use it in any other circumstances.

Notes

1 The method was further developed in the 1970s, and some people distinguish between the original and the new effects method; see Fouirry (1996, p. 241).
2 Value-added is defined in this context in FAO (1986, pp. 45–48). The complete effects method is explained in some manuals, notably Chervel and Le Gall (1976, 1978) and Fabre and Yung (1985).

3 One such demonstration is given in Franck (1996), along with a general discussion of the method.
4 Fouirry (1996, p. 243). Another discussion of the position of the effects method is that in Chervel (1992).
5 Notably the World Bank and the Asian Development Bank.

H

Model answers to readers' worked examples

This appendix presents tentative answers to the readers' questions at the ends of Chapters 2 and 3. Apart from numerical results, there are no unique correct answers, and those given below are merely suggestions.

Questions on Chapter 2 (see page 37)

(a) The presentation of the analysis indicates that the following categories will definitely need separate consideration when it comes to shadow pricing, so they should be separated out in the cost estimates:

- unskilled labour from within the region whose development is to be promoted
- other unskilled labour
- imported machinery
- oil fuel.

In addition to these, and complementing them, it would probably be desirable to separate out:

- skilled national labour
- foreign labour
- local machinery
- imported equipment (such as the turbines and generators)
- local equipment (much of the equipment can probably be made in the country)
- taxes and duties.

The SPFs would probably be something like the following:

- Regional unskilled labour: around 0.4, to favour the region and labour-intensive methods.
- Other unskilled labour: around 0.6, to favour labour-intensive methods.
- Skilled national labour: near 1.0.

- Foreign labour: significantly more than 1.0, to conserve foreign exchange.
- Local machinery: near 1.0, or a little higher to favour labour-intensive methods.
- Imported machinery: significantly more than 1.0, to conserve foreign exchange and favour labour-intensive methods.
- Local equipment: near 1.0, perhaps slightly lower for equipment made in the region.
- Imported equipment: significantly more than 1.0, to conserve foreign exchange.
- Taxes and duties: zero, as they are transfer payments.
- Oil fuel: significantly more than 1.0, to conserve foreign exchange.
- Other: 1.0, as this is a residual category with no particular need for shadow pricing.

In addition to this separation into categories for the purpose of shadow pricing, the cost estimates for distinct elements such as the dam, the power station, the irrigation project and the road works should be kept separate until the last stage of aggregation, to facilitate sensitivity tests, consideration of alternative project packages or purpose-specific financing (Box 4.1, page 81).

(b) Before fixing the SPFs for the analysis, the analyst should consult the national planning office, the treasury, the ministries concerned with the different elements (water, power, agriculture, roads), the regional authorities, and any outside financing agency that is, or might become, involved (World Bank, Inter-American Development Bank). Some of these might have nothing to say on the matter, but it would still be a mistake not to ask them.

(c) In addition to construction costs, the analysis could include, and quantify at least approximately, the following:
  - operation and maintenance costs
  - replacement costs for fast-wearing components
  - land acquisition costs or opportunity costs
  - agricultural production costs on the irrigation scheme (labour, fertilisers, seeds, cultivations)
  - environmental costs or disbenefits, including the lost landscape at the reservoir site
  - development and operating costs for recreational facilities
  - costs of extra healthcare, lost production, ill-health and early death resulting from increased smoking
  - effects of the project's tobacco production on other producers within the region and the country

- costs (if any) associated with the lengthening of the national highway.

(d) Benefits that could be considered include:

- electric energy (distinguishing between peak energy and base load, and between reliable and unreliable energy; the balance changes through the operating period)
- flood attenuation benefits due to the new reservoir
- increased agricultural production on the irrigation project, distinguishing between wheat and tobacco
- visual and environmental benefits from the new reservoir
- benefits (if any) of realigning the national road
- increased satisfaction of smokers (in theory captured in the market price of tobacco, but that may be heavily distorted by taxes and quotas).

(e) The following effects might be excluded from the CBA and mentioned alongside it:

- environmental and visual gains and losses
- effects of realigning the road
- effects on the region
- effects on the country's foreign exchange position
- effects following from the increased supply and reduced price of tobacco.

These are all aspects that have already been listed as possible costs and benefits. In each case a choice must be made, in consultation with government and financing agency people, whether to include the aspect in the CBA or not. Care must be taken to avoid inadvertent double-counting by including an effect in the analysis and then also mentioning it alongside as if it were an addition to the analysis result.

(f) Aspects that are likely to face differing attitudes among different decision-makers and affected people include:

- shadow pricing, especially regarding labour and regional aspects
- environmental and visual impacts
- effects following from the increased supply and reduced price of tobacco
- escalation of the fuel oil price if it is used for valuing hydropower energy by way of a hypothetical thermal alternative.

Questions on Chapter 3 (see page 72)

These model answers generally give all three indicators NPV, B/C and EIRR, for the benefit of readers who wish to practise calculating them; an actual analysis would normally not use all three for sensitivity tests.

(a) The discount rate should be discussed with the planning office or treasury of the Utopian Government, and with the local representative or head office of the Galactic Bank.

(b) The EIRR is 12.8%.

(c) As a quick estimate, the change will reduce most of the benefits (but not those of years 4 to 7) by 10%: say a reduction of PV of all benefits by 9% of MUt$1230, which is about MUt$110; this will reduce the NPV by about MUt$110 to about MUt$367, and the B/C ratio by about 9% to about 1.49. This is a significant but not drastic reduction in these indicators; the EIRR cannot be quickly re-estimated but would change noticeably.

(d) Recalculation shows the revised NPV to be MUt$373, the B/C ratio 1.50, and the EIRR 11.8% instead of 12.8%. The changes are very slightly smaller than the one-minute estimate given to the directors, so there is no need to go out of your way to revise what you told them.

(e) A reduction of all benefits by 20%, keeping the discount rate at 7%, would reduce the NPV from MUt$373 to MUt$148, the B/C ratio from 1.50 to 1.20, and the EIRR from 11.8% to 9.1%. These are very significant reductions, and the cost estimates are sufficiently uncertain to make such a case not unlikely, so the results must be regarded as sensitive to benefit estimates. The *switching value* is a 33% reduction in benefits (such a reduction would bring the B/C ratio at 7% to 1.0 and the EIRR to 7%). It would probably be worthwhile to undertake more detailed studies to improve the precision and reliability of the benefit estimates.

(f) 'This possible earthquake event would reduce the NPV from MUt$373 to MUt$238, the B/C ratio from 1.50 to 1.31, and the EIRR from 11.8% to 10.0%. About 70% of this is due to the delay and the rest to the extra cost in year 3. These are significant reductions in the economic merit of the project, but they do not render it uneconomic and, in view of the relatively small probability of the event, do not represent a major cause for concern. It may be worthwhile at the detailed design stage to study simple ways of reducing the vulnerability to earthquakes during the construction period, but the extra cost of any measures must be balanced against reduced cost-of-risk and may not be justified.' (This insertion into the feasibility study report should leave you safe from severe embarrassment at an enquiry after an earthquake, provided that the estimates of delay and cost proved reasonably realistic.)

(g) Inclusion of the years 21 to 43, with the same annual costs and benefits continuing, would raise the NPV from MUt$373 to MUt$877, B/C from 1.50 to 2.13, and the EIRR from 11.8% to 13.7%. These are significant changes, so the analysis results are sensitive to the choice of analysis period within this range (the NPV and B/C would be much less sensitive at a higher discount rate). At a 7% rate, 20 years is evidently too short an analysis period.

(h) Indicator results for Seizewell E, with replacement of turbines in years 24, 44, 64, etc., discounting at 7%, are as follows:

Analysis period	23 years	43 years	63 years
NPV (MUt$)	490	868	966
B/C	1.65	2.11	2.23
EIRR	12.5%	13.7%	13.8%

The differences are large between 23 and 43 years, but not very large between 43 and 63 years. The analysis should be conducted with a period of at least 43 years so as to give adequate weight to the operating period.

(i) Year 43 is a convenient year to assume the end of production because the turbines which were installed in year 24 reach the end of their life in that year. (Similar reasoning led to the use of the 23-, 43- and 63-year analysis periods above.) The inclusion in the analysis of the decommissioning cost to year 46 and the maintenance costs to year 100 reduces the NPV from MUt$868 to MUt$845 and B/C from 2.11 to 2.05, but the EIRR remains at 13.7% when expressed to one place of decimals. Because of the powerful effect of discounting, even at only 7%, these are small changes, despite the decommissioning cost being of the same order of magnitude as the original capital cost (56% of it, in fact). The effect is almost entirely due to the decommissioning cost, while the assumed maintenance cost makes negligible difference to the indicators, whether assumed to end at year 100 or to continue in perpetuity.

(j) At a 3% per year discount rate the assumptions about decommissioning and maintenance costs after year 43 are more significant than they were at 7%, but still not dramatic; they reduce the NPV by 5.5% and B/C by 13%, the maintenance contributing only 14% to these changes. At a discount rate of 12% they are completely negligible. This highlights the result of high discount rates when significant costs occur late in a project's lifetime.

(k) Ford Focus' challenge, however intemperately expressed, deserves to be taken seriously. It could be said in response that the economic CBA was never intended to be the sole guide to the decision on the implementation of Seizewell E, and it is perfectly proper to consider other guides alongside it, including explicitly ethical ones such as the effect on future generations regardless of timing. However, the decommissioning and subsequent maintenance procedures were designed to high technical standards and were responsibly and carefully costed; the fact that the economic indicators were favourable, especially at low discount rates, means that the power station would, during its 40 years' operation, generate more than enough real surplus resources to cover the decommissioning and maintenance. (Further sensitivity tests show that B/C is favourable at any discount rate from zero to 12%, even with the decommissioning and maintenance costs doubled.) It would be useful if the public enquiry instigated the strongest possible institutional and legislative safeguards to ensure that the resources needed for thorough decommissioning and maintenance really will be earmarked and cannot be siphoned off for other purposes.

I
Worked example

I.1   Introduction

This appendix presents a worked example of a cost–benefit analysis (CBA), based loosely on several actual cases. It serves as an example both of the way the calculation is done and of the way it might be presented in a report. The example is an engineering and agricultural project, namely the rehabilitation of an irrigation scheme, but many aspects would be applicable to other sorts of project. The full presentation would require some dozens of tables, many of which would be sets of similar ones, and for brevity this appendix gives just one table of each such set. To make the text as realistic as possible, as an example for the relevant parts of a feasibility study report, this appendix contains dummy references to further tables and other parts of that imaginary report, in the form *Table?* or *Annex?*; those annexes and tables do not exist in this book. The rest of this appendix is thus a simplified model of one annex in the feasibility study report, prepared by Lear Consultants for the Amnesian Ministry of Agriculture, on the rehabilitation project. Some of the tables are similar to those in Section 4.2 of this book, which explains them more fully than this text on an imitation feasibility report can.

Although mainly devoted to an economic CBA aimed at guiding the decision of whether or not to implement the rehabilitation project, this appendix also contains a farm-level financial analysis and some remarks about its significance as a determinant of project viability. A project assessment often needs to contain several such single-party financial analyses, to prove the motivation of the parties and thus the viability of the project.

I.2   The project

The Upper Gromboolia Irrigation Scheme (UGIS) was constructed between 1972 and 1977, providing irrigation to a previously rain-fed area of the Great Gromboolian Plain by means of a system of canals fed from reservoirs formed by dams on three rivers in the Hills of the Chankly Bore, in the People's Republic of Amnesia. The purpose of the scheme is to increase agricultural production for both local consumption and import substitution, with the following four farming systems:

- smallholder irrigation by local people, growing maize, cowpeas and some vegetables on farms of typical size, 5 ha
- large arable farms operated by farmers with considerable capital resources
- rangeland, on the soils less suited to arable cropping, used for grazing of cattle and goats
- forestry and woodlots on selected patches of land, mostly for fuel wood.

Agricultural production has declined in recent years due to general wear and tear, insufficient maintenance, inadequate roads, sedimentation of the reservoirs, and suboptimal operation methods resulting in significant wastage of water.

The purpose of the Upper Gromboolia Irrigation Rehabilitation Project (UGIRP) is to rehabilitate and upgrade the scheme's irrigation and drainage infrastructure, and its internal roads, in order to raise agricultural production. Project elements include lining of some canals, heightening of two of the dams, improvements to drains and roads, new structures for control and measuring of water flows, improved operating methods, and enhanced staffing and staff training. These project elements are described fully, with cost estimates, in *annex?*.

This annex presents the economic CBA of the rehabilitation project. It is conducted with the domestic pricing numéraire, using border prices and a shadow exchange rate for traded goods. The currency used in the calculation is Amnesian pesos at mid-1998 prices. The official exchange rate for that price datum is 2.53 pesos per US dollar. After discussion with the Ministry of Planning, it has been agreed to use a discount rate for economic analysis of 8% per year, and a shadow exchange rate factor of 1.15, which implies a shadow exchange rate of 2.91 pesos per dollar. The analysis period is 30 years from the start of the construction of the rehabilitation works. Agricultural production costs are deducted from gross crop returns to give net crop returns, and the incremental net returns are used to calculate benefits; thus agricultural production costs are treated in this analysis as negative benefits and are not included in the project costs, which comprise all operations

242

carried out by parties other than the farmers (mainly the scheme administration but also the extension and training divisions of the Ministry of Agriculture).

The with-project situation is defined as the full implementation of the project as described in *annex?*, with prices and crop yields as shown in this annex. The without-project situation assumes that agricultural production will be maintained at its present level, without further decline, by operation and maintenance efforts sufficient for that purpose: the economic cost of those efforts is included in the estimation of the incremental cost of the rehabilitation project.

I.3   Economic prices

The detailed estimation of economic prices is presented in *annex?*, from which Tables 1 and 2 in this section are reproduced.

Table 1 Basic parameters and prices

All in pesos at 1998 constant prices, domestic pricing system
Official exchange rate   2.530 pesos/US$
Shadow exchange rate factor   1.150
Shadow exchange rate   2.910 pesos/US$
Discount rate   8%/year

Good	Unit	Financial price	Economic price	Shadow price factor
Investment inputs				
Imported materials				1.10
Other materials				1.00
Skilled labour				1.10
Unskilled labour				0.60
Machine use				1.20
Agricultural inputs				
Maize seed	kg	0.2	0.2	
Fertiliser:				
N	kg	2.86	3.07	
P_2O_5	kg	2.30	2.43	
K_2O	kg	2.39	2.54	
Tractor use	hour	25.00	28.00	
Hired labour	person-day	12.00	5.00	
Family labour	person-day	0.00	5.00	
Outputs				
Maize	t	565.0	544.0	
Cabbage	t	895.0	925.0	
Tomatoes	t	805.0	815.0	
Cowpeas	t	640.0	655.0	

Table 2 Economic price of maize, as import substitute

Unit: tonne (shadow price factors (SPFs) for handling and transport are weighted means according to tax/traded/non-traded proportions)		Pesos at 1998 prices		
		Financial price	SPF	Economic price
World price of maize in US$				
World Bank forecast for 1998, US No. 2 maize fob US Gulf port	US$	116.00		
Quality adjustment factor, project/US No. 2	factor	1.04		
Adjusted price for project maize	US$	120.64		
Freight, US Gulf port to Amnesian port	US$	53.00		
Insurance, etc., to Amnesian port	US$	8.00		
CIF price, imported maize, Amnesian port	US$	181.64		
Border equivalent price in local currency				
Exchange rates		2.530		2.910
CIF price, imported maize, Amnesian port	pesos	459.55		528.48
Import and other taxes	pesos	89.25	zero	0.00
Port handling costs	pesos	10.20	1.05	10.71
Transport, port to Fort Edward market	pesos	22.00	0.80	17.60
Deduct: transport, project to market	pesos	−16.00	0.80	−12.80
Border equivalent farm gate price	**pesos**	**565.00**		**543.99**

Table 1 brings together the economic prices and shadow price factors of all the significant categories and goods, details being in *annex?* Rural unemployment is severe on the Great Gromboolian Plain, and the economic prices of unskilled labour for construction and of agricultural labour reflect this. In the light of seasonal factors, and after discussion with the Ministries of Finance and of Agriculture, the values assumed are an economic price of agricultural labour, including unpaid family labour on smallholder farms, of 5.00 pesos per person-day (the wage to hired agricultural labour being 12.00 pesos), and a shadow price factor (SPF) of 0.60 for all other unskilled labour.

The last section in Table 1 gives the financial and economic prices of the project's outputs. Table 2 shows the derivation of the economic price of maize at the project site, corresponding tables for the other crops being in *annex?* Maize produced on this project will serve to reduce the country's total maize imports, as Amnesia is expected to continue to import maize throughout the analysis period. The forecast world price of maize is US$116 per tonne, free-on-board (fob) US

Gulf port. This is for maize of grade US No. 2, and the maize produced by the project is expected to be of a slightly higher quality, so a quality premium is applied in the form of an upward adjustment of 4%. Adding the carriage, insurance and freight (CIF) costs of bringing the imported maize to the border gives a *border price* of US$181.64 per tonne at the Amnesian port. This CIF border price is then converted into local currency at the official exchange rate for financial prices, and at the shadow exchange rate for the economic analysis. As the place of substitution of project maize for imported maize is the wholesale market in Fort Edward, the cost of transporting the imported maize from the border to that market is added, along with taxes and costs at the border, to obtain the border equivalent price at Fort Edward. To obtain the border equivalent price at the project site (farm gate price), the costs of transporting project-produced maize from the site to Fort Edward is subtracted. The SPFs of the handling and transport costs have been estimated from their approximate breakdown into traded and non-traded components. The resulting economic farm-gate value of maize at 1998 prices is 544 pesos/t, while the financial price is 565 pesos/t.

I.4   Farm level financial analysis

The crop budget for smallholder-grown irrigated maize in the with-project situation is given in financial prices in Table 3. This shows that maize gives a financial return, omitting any payments for land or water, of 1392 pesos per hectare of land and 23 pesos/person-day of family labour. This return is expected to ensure that smallholder farming families will choose to grow maize on about 60% of their land. Vegetables are represented for the purposes of analysis by cabbages and tomatoes, although small amounts of others will be grown; they give higher returns to land than maize, but have high labour requirements and some market limitations, so they are assumed to be grown on only 20% of the land, with cowpeas for local consumption on the remaining 20%. The financial farm budget for a typical 5 ha smallholder farm with this cropping pattern is shown in Table 4 (tables for other crops and farming systems, and for the without-project situation, are in *annex?*). This table indicates that such a farm is expected to show a financial return, in the absence of payments for land, water or family labour, of some 7146 pesos/year. With an annual input for the 5 ha farm of 329 person-days, this represents an average return on family labour of 22 pesos/person-day, which is nearly twice the agricultural

Table 3 Financial crop budget, per hectare

Crop	Maize			
Farming system	Smallholder irrigation			
Situation	Future with-project			
Prices	Financial			
Currency	Pesos at 1998 constant prices			

Item	Unit	Quantity	Price	Value
Gross return				
Main product	t	4.0	565	2260
By-product	t			0
Losses	t			0
			Gross return	2260
Production costs				
Seed	kg	20	0.20	4
Fertiliser:				
N	kg	110.0	2.86	315
P_2O_5	kg	60.0	2.30	138
K_2O	kg	90.0	2.39	215
Agrochemicals	peso	40.0		40
Machinery:				
Tractor	h	2.0	25.00	50
Combine	h	0.0		0
Animal power	peso	10.0		10
Sacks, etc.	peso	0.0		0
Hired labour	person-day	8.0	12.00	96
Family labour	person-day	60.0	0.00	0
			Total costs	868
Net economic return per hectare		Pesos per hectare		1392
Net return to family labour		Pesos per person-day		23

daily wage rate of 12 pesos. Some farm families are tenants and have to pay up to 2000 pesos/year for the use of the land, but even in that case there would be sufficient residual return for farmers to pay a water charge of up to 1200 pesos/year (240 pesos per hectare per year) before the return per family person-day would dip below the hired labour rate of 12 pesos/day. Alternative water charge scenarios, with charge rates of 50–200 pesos per hectare per year, are used in the cost recovery analysis for the scheme operating authority in *annex?*

I.5   Economic benefits

Table 5 gives the economic crop budget for smallholder maize in the with-project situation, corresponding to the financial crop budget in Table 3. The economic farm budget for smallholders in the with-project

Table 4 Financial farm budget

Farming system	Smallholder irrigation
Situation	Future with-project
Prices	Financial
Currency	Pesos at 1998 constant prices

Crop	Area for this crop: ha	Net return: pesos		Family labour	
		Per hectare	Per farm	Per hectare	Per farm
Maize	3.0	1392	4177	60	180
Cabbage	0.5	2270	1135	110	55
Tomatoes	0.5	2630	1315	125	63
Cowpeas	1.0	519	519	31	31
Total	5.0		7146		329
Farm net return per hectare			1429		
Farm net return per person-day of family labour					22

Table 5 Economic crop budget, per hectare

Crop	Maize
Farming system	Smallholder irrigation
Situation	Future with-project
Prices	Economic
Currency	Pesos at 1998 constant prices

Item	Unit	Quantity	Price	Value
Gross return				
Main product	t	4.0	544	2176
By-product	t			0
Losses	t			0
			Gross return	2176
Production costs				
Seed	kg	20.0	0.20	4
Fertiliser:				
N	kg	110.0	3.07	338
P_2O_5	kg	60.0	2.43	146
K_2O	kg	90.0	2.54	229
Agrochemicals	peso	40.0		40
Machinery:				
Tractor	h	2.0	28.00	56
Combine	h	0.0		0
Animal power	peso	10.0		10
Sacks, etc.	peso	0.0		0
Hired labour	person-day	8.0	5.00	40
Family labour	person-day	60.0	5.00	300
			Total costs	1162
Net economic return		Pesos per hectare		1014

Table 6 Economic farm budget

Farming system	Smallholder irrigation
Situation	Future with-project
Prices	Economic
Currency	Pesos at 1998 constant prices

Crop	Area for this crop: ha	Net return: pesos	
		Per hectare	Per farm
Maize	3.0	1014	3042
Cabbage	0.5	2045	1023
Tomatoes	0.5	2427	1214
Cowpeas	1.0	495	495
Total	5.0		5773
Farm net return per hectare			1155

situation is given in Table 6, and the tables for other crops and farming systems, and for the without-project situation, are in *annex?* Table 7 brings the resulting farm net returns together for the whole scheme, the net cultivable area of which is assumed to be divided between the farming systems as shown in the table, with 3700 ha occupied by smallholder farms. The incremental economic benefit is given by the differences between with-project and without-project situations, and it can be seen that the rehabilitation is expected to increase the scheme's overall economic return from 1.64 million to 8.22 million pesos/year. Just over half the incremental economic return comes from the smallholder irrigation farming system. Most of the increase is

Table 7 Project incremental benefits per year in operation phase

Farming system	Area used: ha	Without-project economic return		With-project economic return		Incremental economic benefit	
		Per hectare: pesos/year	Total: kilo-pesos/ year	Per hectare: pesos/year	Total: kilo-pesos/ year	Total: kilo-pesos/ year	%
Smallholder irrigation	3700	247	914	1155	4272	3358	51
Large arable	2050	196	402	1258	2579	2177	33
Rangeland	1370	108	148	470	644	496	8
Forest/ woodlots	1860	95	177	389	724	547	8
Total	8980		1640		8218	6578	

Table 8 Build-up of incremental benefits

Proportion of full incremental benefits: %	Project year	Economic benefits: million pesos
0	1	0.00
0	2	0.00
10	3	0.66
25	4	1.64
45	5	2.96
65	6	4.28
85	7	5.59
95	8	6.25
100	9	6.58
	10 to 30	6.58

due to a higher proportion of wheat in the cropping pattern and to improved yields, both resulting from the larger quantity of irrigation water and its much improved distribution.

The rehabilitation work is to be done in stages over three years, and it is expected to take several years for the full incremental benefit to be achieved, largely because of the need for smallholder farmers to become familiar with the better water management and modern use of fertilisers that are needed to mobilise the benefits of the extra irrigation water. The assumed rate of build-up of incremental economic benefits at project level is shown in Table 8.

I.6 Economic costs

The derivation of the economic costs of the project works, including periodic replacement of water control equipment, is given in Table 9. The starting point is the financial cost estimates and breakdowns for rehabilitation work in years 1, 2, 3 and replacements in year 16, all taken from *annex?* These are multiplied by the shadow price factors (SPFs) from Table 1 to obtain the economic costs. Taxes and duties, being transfer payments, do not constitute economic costs. There are no corresponding costs in the without-project situation.

The annual operating costs in both with-project and without-project situations are given in Table 10, details being in *annex?*, their difference giving the incremental annual operating cost. The estimates include extension services and training. The without-project case involves significant costs to prevent a further decline in agricultural production, but the with-project situation involves rather higher costs because of the higher standard of operation and maintenance assumed, supported

Table 9 Construction and replacement costs

		Million pesos at 1998 prices		
		Financial prices	SPF	Economic prices
Year 1				
Taxes and duties		0.23	–	0.00
Imported materials		1.07	1.10	1.18
Other materials		1.35	1.00	1.35
Skilled labour		1.29	1.10	1.42
Unskilled labour		1.71	0.60	1.03
Machine use		2.34	1.20	2.81
	Total	7.99		7.78
Year 2				
Taxes and duties		0.86	–	0.00
Imported materials		2.19	1.10	2.41
Other materials		3.10	1.00	3.10
Skilled labour		4.13	1.10	4.54
Unskilled labour		5.02	0.60	3.01
Machine use		4.87	1.20	5.84
	Total	20.17		18.91
Year 3				
Taxes and duties		0.34	–	0.00
Imported materials		1.67	1.10	1.84
Other materials		0.96	1.00	0.96
Skilled labour		1.80	1.10	1.98
Unskilled labour		1.47	0.60	0.88
Machine use		2.89	1.20	3.47
	Total	9.13		9.13
Year 4				
Taxes and duties		0.80	–	0.00
Imported materials		1.03	1.10	1.13
Other materials		0.04	1.00	0.04
Skilled labour		0.78	1.10	0.86
Unskilled labour		0.85	0.60	0.51
Machine use		0.43	1.20	0.52
	Total	3.93		3.06

by training and improved extension services. The shadow pricing is similar to that for construction costs, and the incremental annual economic cost is estimated at 980 000 pesos.

I.7   Cost–benefit indicators

The incremental economic costs and benefits from Tables 8 to 10 are combined in Table 11 to calculate the rehabilitation project's net benefit in each year of the analysis period. The indicator values are:

Table 10 Project incremental operating costs per year

	Million pesos at 1998 prices		
	Financial prices	SPF	Economic prices
Without-project situation			
Taxes and duties	0.23	–	0.00
Imported materials	0.11	1.10	0.12
Other materials	0.13	1.00	0.13
Skilled labour	0.43	1.10	0.47
Unskilled labour	0.71	0.60	0.43
Machine use	0.30	1.20	0.36
Total	1.91		1.51
With-project situation			
Taxes and duties	0.41	–	0.00
Imported materials	0.19	1.10	0.21
Other materials	0.28	1.00	0.28
Skilled labour	0.65	1.10	0.72
Unskilled labour	1.08	0.60	0.65
Machine use	0.53	1.20	0.64
Total	3.14		2.49
Incremental cost	1.23		0.98

- net present value at 8% discount rate: 9.3 million pesos at 1998 economic prices
- benefit/cost ratio at 8% discount rate: 1.23
- economic internal rate of return (EIRR): 10.5%.

I.8 Sensitivity tests

A number of *sensitivity tests* have been carried out to show which estimates are critical in determining the indicator values. The results for the B/C ratio are given in Table 12. The sensitivity indicator is the approximate percentage change in the B/C ratio for a 1% change in the parameter that is varied in a particular sensitivity test. The switching value is the value of the parameter, as a percentage of its base-case value, that would bring the B/C ratio to 1.0, other parameters remaining at their base case values.

These sensitivity tests indicate that the B/C ratio is fairly sensitive to construction costs and to smallholder irrigated maize yields, changing by around 20% for a 30% change in either of these parameters (sensitivity index around 0.7). It is also quite sensitive to the forecast world price of maize, dropping by over 14% when that price drops

Table 11 Economic costs and benefits by year

In 1998 economic pesos, using domestic pricing numéraire, with official exchange rate 2.530 pesos/US$ and shadow exchange rate 2.910 pesos/US$

Year	Costs				Benefits	Net benefits
	Construction	Replace	Operating	Total		
1	7.78			7.78		−7.78
2	18.91		0.98	18.91		−18.91
3	9.13		0.98	9.13	0.66	−8.47
4				0.98	1.64	0.67
5				0.98	2.96	1.98
6			0.98	0.98	4.28	3.30
7			0.98	0.98	5.59	4.61
8			0.98	0.98	6.25	5.27
9			0.98	0.98	6.58	5.60
10			0.98	0.98	6.58	5.60
11			0.98	0.98	6.58	5.60
12			0.98	0.98	6.58	5.60
13			0.98	0.98	6.58	5.60
14			0.98	0.98	6.58	5.60
15			0.98	0.98	6.58	5.60
16		3.06	0.98	4.04	6.58	2.54
17			0.98	0.98	6.58	5.60
18			0.98	0.98	6.58	5.60
19			0.98	0.98	6.58	5.60
20			0.98	0.98	6.58	5.60
21			0.98	0.98	6.58	5.60
22			0.98	0.98	6.58	5.60
23			0.98	0.98	6.58	5.60
24			0.98	0.98	6.58	5.60
25			0.98	0.98	6.58	5.60
26			0.98	0.98	6.58	5.60
27			0.98	0.98	6.58	5.60
28			0.98	0.98	6.58	5.60
29			0.98	0.98	6.58	5.60
30			0.98	0.98	6.58	5.60
At discount rate 8%/year	Present values			40.04	49.33	9.29
	B/C ratio 1.232					

Economic internal rate of return 10.51

30%. Another important estimate is the timing of benefit start-up: a one-year delay would lower the B/C ratio considerably, and a two-year delay would bring it below 1.0 and render the project uneconomic, other things being equal. The B/C ratio is, however, only mildly

Table 12 Sensitivity tests for UGIRP economic analysis

Test details	B/C	Sensitivity indicator	Switching value
Base case (as in Tables 1 to 11)	1.232		
Construction costs 30% higher	1.002	−0.6	130%
Operating costs 30% higher	1.158	−0.2	209%
SPF of unskilled construction labour 0.80 instead of 0.60 (up by a third)	1.170	−0.15	SPF of 1.5
Maize world price 20% lower	1.115	+0.5	US$70/t
Smallholder maize yield 20% lower	1.051	+0.7	2.97 t/ha
Economic net return on smallholder irrigated vegetables 20% lower	1.170	+0.25	25% (75% down)
Economic value of agricultural labour 10.00 pesos instead of 5.00 pesos per person per day (up by 100%)	1.091	−0.1	13.20 pesos per person per day
Smallholder irrigated with-project cropping pattern: maize proportion decreased by a third to 2 ha out of 5 ha (area switched to cowpeas)	1.160	+0.2	none (B/C = 1.02 even with no maize)
Project's area under smallholders decreased by 500 ha (or 13.5%) to 3200 ha (switched to large arable)	1.247	−0.1	none
Benefit start delayed by one year	1.126	−	1.9 years
Benefit start delayed by two years	0.990		

sensitive to the other parameters tested, with sensitivity indicator values of 0.1 to 0.25.

I.9   Conclusion

The result of this economic CBA is that the rehabilitation project as currently formulated is economically acceptable by the criterion of a test discount rate of 8%, but by a fairly small margin: the benefit/cost ratio at that rate is only 1.23 and the economic EIRR is 10.5%/year.

Sensitivity tests show that this result is fairly sensitive to forecasts of initial cost and of the yield, and the world price of the wheat grown by smallholders under irrigation. It is not particularly sensitive to assumptions about cropping patterns and land tenure, the growing of vegetables, or the value of agricultural labour. The tests indicate that the economic indicators could be slightly higher if more prominence were given to large arable farms, or to vegetables grown by smallholders, but the former would be socially unsatisfactory and the latter unrealistic in the light of market opportunities for vegetables, and of limited farm labour.

The financial analysis for smallholder farms, described in Section I.4, shows that farm families can be expected to be motivated to engage in irrigated agriculture with approximately the cropping pattern assumed, and that they would be able to pay water charges of the order of 100 pesos/ha per year without losing that motivation. The financial analysis of the scheme operating authority (in *annex?*, not summarised here) shows that the authority could operate without subsidy if it levied such a water charge from all farmers. The rehabilitation project thus appears to be financially viable at both these levels.

This result must be compared with other projects currently available for possible implementation in the country. Alongside this economic analysis result, the Ministry is recommended to consider the generally favourable conclusions of the social and environmental investigations reported in *annex?*

I.10   Postscript

As explained in Section I.1, the intervening sections of this appendix have presented, in abbreviated form and with many references to other parts of an imaginary feasibility study, the economic analysis of a rehabilitation project, along with a financial viability analysis for one of the groups involved. Readers who require practice with the techniques of CBA can use this worked example in a number of ways, such as:

- checking the arithmetic – it is always a good idea to run spot checks on any analysis, especially one done by computer, to detect errors due to incorrect copying of figures from one place to another, or adding or discounting over the wrong ranges (the account given above is designed to be totally transparent, so the reader should find all necessary parameter values in the tables or text)
- reproducing the whole analysis in a spreadsheet (which may be the easiest way of checking it; Tables 1 to 11 above were all calculated and printed using one spreadsheet; small differences should not cause concern as they are likely to be caused by rounding of numbers)
- designing and conducting additional sensitivity tests, for instance on the proportion of vegetables in the smallholder cropping pattern
- considering ways to improve the reporting of the analysis (perhaps with reference to the checklist in Chapter 6); perhaps more should have been said about the farm-level financial returns without the rehabilitation project.

Some of the tables shown in this sample report could have been omitted or simplified, but it is often worthwhile to present intermediate numbers like those in Tables 6 and 11 so as to help readers to follow the analysis and to conduct their own sensitivity tests. Some intermediate numbers are shown in tables to more significant figures than their precision warrants, in order to clarify the analysis, but the final results are rounded when quoted in the text. The B/C ratios are given to three decimal places in Table 12 only because the sensitivity tests involve relatively small differences between them: these differences are significant even though the absolute B/C values are, of course, not precise to that level.

J
Discounting tables

Many readers will routinely use spreadsheet or workbook software and a computer for discounting, and this is explained in Chapter 3, Box 3.3, page 50. The tables in this appendix are included for those who need them, and also to help readers to work through the examples in the text and to get a feel for the effect and strength of discounting. It is wise occasionally to check figures produced by a computer: the software is error-free but the user may have made a mistake, especially in the definition of a range. The tables that follow give values of discounting factors PV/FV and PV/RV, which are defined and explained in Section 3.2.

- PV means present value (by convention always in year 0)
- FV means a future value in year n
- RV means the value of each of a series of future values in years 1 to n.

The appendix presents 36 separate tables, each for one discount rate, which is shown at the top left corner of the table. Each table has a column for the number of years n and columns for the two factors. There is a table for every whole-number percentage discount rate up to 20%, then every second one up to 36%, then every fifth from 40% to 75%, although such high rates are very seldom applicable; the factors for intermediate rates can be found to an adequate level of precision by linear interpolation.

For the smaller discount rates the tables give values for every year up to 50 years and then every fifth year up to 100 years; again the factors for intermediate periods can be arrived at by linear interpolation. The factors are given correct to four decimal places, except for some PV/FV factors less than about 0.1 which are given to five places. For larger discount rates some of the larger values of n are omitted because their factors expressed to four or five places of decimals are constant and hence obvious, although the values for 100 years are always given.

Cost–benefit analysis – A practical guide
ISBN: 978-0-7277-4134-9

A reader using the tables does not need to know the formulae for the two functions. For those wishing to compute a factor directly, the formulae, for discount rate r (as a ratio, not a percentage) are:

$$PV/FV_{(year\, n)} = \frac{1}{(1+r)^n} \qquad \text{(as given in Section 3.2)}$$

which for large n and large r tends to zero, and

$$PV/FV_{(year\, 1\, to\, n)} = \frac{1}{r}\left\{1 - \frac{1}{(1+r)^n}\right\}$$

which for large n, or for a series to perpetuity, tends to $1/r$; for any rate over about 12% the PV/RV factor for 100 years is effectively the same as that for perpetuity.

Discounting tables, 1% to 4%

1%/year rate			2%/year rate			3%/year rate			4%/year rate		
No. of years	PV/FV	PV/RV	No. of years	PV/FV	PV/RV	No. of years	PV/FV	PV/RV	No. of years	PV/FV	PV/RV
1	0.9901	0.9901	1	0.9804	0.9804	1	0.9709	0.9709	1	0.9615	0.9615
2	0.9803	1.9704	2	0.9612	1.9416	2	0.9426	1.9135	2	0.9246	1.8861
3	0.9706	2.9410	3	0.9423	2.8839	3	0.9151	2.8286	3	0.8890	2.7751
4	0.9610	3.9020	4	0.9238	3.8077	4	0.8885	3.7171	4	0.8548	3.6299
5	0.9515	4.8534	5	0.9057	4.7135	5	0.8626	4.5797	5	0.8219	4.4518
6	0.9420	5.7955	6	0.8880	5.6014	6	0.8375	5.4172	6	0.7903	5.2421
7	0.9327	6.7282	7	0.8706	6.4720	7	0.8131	6.2303	7	0.7599	6.0021
8	0.9235	7.6517	8	0.8535	7.3255	8	0.7894	7.0197	8	0.7307	6.7327
9	0.9143	8.5660	9	0.8368	8.1622	9	0.7664	7.7861	9	0.7026	7.4353
10	0.9053	9.4713	10	0.8203	8.9826	10	0.7441	8.5302	10	0.6756	8.1109
11	0.8963	10.3676	11	0.8043	9.7868	11	0.7224	9.2526	11	0.6496	8.7605
12	0.8874	11.2551	12	0.7885	10.5753	12	0.7014	9.9540	12	0.6246	9.3851
13	0.8787	12.1337	13	0.7730	11.3484	13	0.6810	10.6350	13	0.6006	9.9856
14	0.8700	13.0037	14	0.7579	12.1062	14	0.6611	11.2961	14	0.5775	10.5631
15	0.8613	13.8651	15	0.7430	12.8493	15	0.6419	11.9379	15	0.5553	11.1184
16	0.8528	14.7179	16	0.7284	13.5777	16	0.6232	12.5611	16	0.5339	11.6523
17	0.8444	15.5623	17	0.7142	14.2919	17	0.6050	13.1661	17	0.5134	12.1657
18	0.8360	16.3983	18	0.7002	14.9920	18	0.5874	13.7535	18	0.4936	12.6593
19	0.8277	17.2260	19	0.6864	15.6785	19	0.5703	14.3238	19	0.4746	13.1339
20	0.8195	18.0456	20	0.6730	16.3514	20	0.5537	14.8775	20	0.4564	13.5903
21	0.8114	18.8570	21	0.6598	17.0112	21	0.5375	15.4150	21	0.4388	14.0292
22	0.8034	19.6604	22	0.6468	17.6580	22	0.5219	15.9369	22	0.4220	14.4511
23	0.7954	20.4558	23	0.6342	18.2922	23	0.5067	16.4436	23	0.4057	14.8568
24	0.7876	21.2434	24	0.6217	18.9139	24	0.4919	16.9355	24	0.3901	15.2470
25	0.7798	22.0232	25	0.6095	19.5235	25	0.4776	17.4131	25	0.3751	15.6221
26	0.7720	22.7952	26	0.5976	20.1210	26	0.4637	17.8768	26	0.3607	15.9828
27	0.7644	23.5596	27	0.5859	20.7069	27	0.4502	18.3270	27	0.3468	16.3296
28	0.7568	24.3164	28	0.5744	21.2813	28	0.4371	18.7641	28	0.3335	16.6631
29	0.7493	25.0658	29	0.5631	21.8444	29	0.4243	19.1885	29	0.3207	16.9837
30	0.7419	25.8077	30	0.5521	22.3965	30	0.4120	19.6004	30	0.3083	17.2920
31	0.7346	26.5423	31	0.5412	22.9377	31	0.4000	20.0004	31	0.2965	17.5885
32	0.7273	27.2696	32	0.5306	23.4683	32	0.3883	20.3888	32	0.2851	17.8736
33	0.7201	27.9897	33	0.5202	23.9886	33	0.3770	20.7658	33	0.2741	18.1476
34	0.7130	28.7027	34	0.5100	24.4986	34	0.3660	21.1318	34	0.2636	18.4112
35	0.7059	29.4086	35	0.5000	24.9986	35	0.3554	21.4872	35	0.2534	18.6646
36	0.6989	30.1075	36	0.4902	25.4888	36	0.3450	21.8323	36	0.2437	18.9083
37	0.6920	30.7995	37	0.4806	25.9695	37	0.3350	22.1672	37	0.2343	19.1426
38	0.6852	31.4847	38	0.4712	26.4406	38	0.3252	22.4925	38	0.2253	19.3679
39	0.6784	32.1630	39	0.4619	26.9026	39	0.3158	22.8082	39	0.2166	19.5845
40	0.6717	32.8347	40	0.4529	27.3555	40	0.3066	23.1148	40	0.2083	19.7928
41	0.6650	33.4997	41	0.4440	27.7995	41	0.2976	23.4124	41	0.2003	19.9931
42	0.6584	34.1581	42	0.4353	28.2348	42	0.2890	23.7014	42	0.1926	20.1856
43	0.6519	34.8100	43	0.4268	28.6616	43	0.2805	23.9819	43	0.1852	20.3708
44	0.6454	35.4555	44	0.4184	29.0800	44	0.2724	24.2543	44	0.1780	20.5488
45	0.6391	36.0945	45	0.4102	29.4902	45	0.2644	24.5187	45	0.1712	20.7200
46	0.6327	36.7272	46	0.4022	29.8923	46	0.2567	24.7754	46	0.1646	20.8847
47	0.6265	37.3537	47	0.3943	30.2866	47	0.2493	25.0247	47	0.1583	21.0429
48	0.6203	37.9740	48	0.3865	30.6731	48	0.2420	25.2667	48	0.1522	21.1951
49	0.6141	38.5881	49	0.3790	31.0521	49	0.2350	25.5017	49	0.1463	21.3415
50	0.6080	39.1961	50	0.3715	31.4236	50	0.2281	25.7298	50	0.1407	21.4822
55	0.5785	42.1472	55	0.3365	33.1748	55	0.1968	26.7744	55	0.1157	22.1086
60	0.5504	44.9550	60	0.3048	34.7609	60	0.1697	27.6756	60	0.0951	22.6235
65	0.5237	47.6266	65	0.2761	36.1975	65	0.1464	28.4529	65	0.07813	23.0467
70	0.4983	50.1685	70	0.2500	37.4986	70	0.1263	29.1234	70	0.06422	23.3945
75	0.4741	52.5871	75	0.2265	38.6771	75	0.1089	29.7018	75	0.05278	23.6804
80	0.4511	54.8882	80	0.2051	39.7445	80	0.0940	30.2008	80	0.04338	23.9154
85	0.4292	57.0777	85	0.1858	40.7113	85	0.08107	30.6312	85	0.03566	24.1085
90	0.4084	59.1609	90	0.1683	41.5869	90	0.06993	31.0024	90	0.02931	24.2673
95	0.3886	61.1430	95	0.1524	42.3800	95	0.06032	31.3227	95	0.02409	24.3978
100	0.3697	63.0289	100	0.1380	43.0984	100	0.05203	31.5989	100	0.01980	24.5050

Discounting tables, 5% to 8%

5%/year rate			6%/year rate			7%/year rate			8%/year rate		
No. of years	PV/FV	PV/RV	No. of years	PV/FV	PV/RV	No. of years	PV/FV	PV/RV	No. of years	PV/FV	PV/RV
1	0.9524	0.9524	1	0.9434	0.9434	1	0.9346	0.9346	1	0.9259	0.9259
2	0.9070	1.8594	2	0.8900	1.8334	2	0.8734	1.8080	2	0.8573	1.7833
3	0.8638	2.7232	3	0.8396	2.6730	3	0.8163	2.6243	3	0.7938	2.5771
4	0.8227	3.5460	4	0.7921	3.4651	4	0.7629	3.3872	4	0.7350	3.3121
5	0.7835	4.3295	5	0.7473	4.2124	5	0.7130	4.1002	5	0.6806	3.9927
6	0.7462	5.0757	6	0.7050	4.9173	6	0.6663	4.7665	6	0.6302	4.6229
7	0.7107	5.7864	7	0.6651	5.5824	7	0.6227	5.3893	7	0.5835	5.2064
8	0.6768	6.4632	8	0.6274	6.2098	8	0.5820	5.9713	8	0.5403	5.7466
9	0.6446	7.1078	9	0.5919	6.8017	9	0.5439	6.5152	9	0.5002	6.2469
10	0.6139	7.7217	10	0.5584	7.3601	10	0.5083	7.0236	10	0.4632	6.7101
11	0.5847	8.3064	11	0.5268	7.8869	11	0.4751	7.4987	11	0.4289	7.1390
12	0.5568	8.8633	12	0.4970	8.3838	12	0.4440	7.9427	12	0.3971	7.5361
13	0.5303	9.3936	13	0.4688	8.8527	13	0.4150	8.3577	13	0.3677	7.9038
14	0.5051	9.8986	14	0.4423	9.2950	14	0.3878	8.7455	14	0.3405	8.2442
15	0.4810	10.3797	15	0.4173	9.7122	15	0.3624	9.1079	15	0.3152	8.5595
16	0.4581	10.8378	16	0.3936	10.1059	16	0.3387	9.4466	16	0.2919	8.8514
17	0.4363	11.2741	17	0.3714	10.4773	17	0.3166	9.7632	17	0.2703	9.1216
18	0.4155	11.6896	18	0.3503	10.8276	18	0.2959	10.0591	18	0.2502	9.3719
19	0.3957	12.0853	19	0.3305	11.1581	19	0.2765	10.3356	19	0.2317	9.6036
20	0.3769	12.4622	20	0.3118	11.4699	20	0.2584	10.5940	20	0.2145	9.8181
21	0.3589	12.8212	21	0.2942	11.7641	21	0.2415	10.8355	21	0.1987	10.0168
22	0.3418	13.1630	22	0.2775	12.0416	22	0.2257	11.0612	22	0.1839	10.2007
23	0.3256	13.4886	23	0.2618	12.3034	23	0.2109	11.2722	23	0.1703	10.3711
24	0.3101	13.7986	24	0.2470	12.5504	24	0.1971	11.4693	24	0.1577	10.5288
25	0.2953	14.0939	25	0.2330	12.7834	25	0.1842	11.6536	25	0.1460	10.6748
26	0.2812	14.3752	26	0.2198	13.0032	26	0.1722	11.8258	26	0.1352	10.8100
27	0.2678	14.6430	27	0.2074	13.2105	27	0.1609	11.9867	27	0.1252	10.9352
28	0.2551	14.8981	28	0.1956	13.4062	28	0.1504	12.1371	28	0.1159	11.0511
29	0.2429	15.1411	29	0.1846	13.5907	29	0.1406	12.2777	29	0.1073	11.1584
30	0.2314	15.3725	30	0.1741	13.7648	30	0.1314	12.4090	30	0.0994	11.2578
31	0.2204	15.5928	31	0.1643	13.9291	31	0.1228	12.5318	31	0.0920	11.3498
32	0.2099	15.8027	32	0.1550	14.0840	32	0.1147	12.6466	32	0.0852	11.4350
33	0.1999	16.0025	33	0.1462	14.2302	33	0.1072	12.7538	33	0.0789	11.5139
34	0.1904	16.1929	34	0.1379	14.3681	34	0.1002	12.8540	34	0.0730	11.5869
35	0.1813	16.3742	35	0.1301	14.4982	35	0.0937	12.9477	35	0.0676	11.6546
36	0.1727	16.5469	36	0.1227	14.6210	36	0.0875	13.0352	36	0.0626	11.7172
37	0.1644	16.7113	37	0.1158	14.7368	37	0.0818	13.1170	37	0.0580	11.7752
38	0.1566	16.8679	38	0.1092	14.8460	38	0.0765	13.1935	38	0.0537	11.8289
39	0.1491	17.0170	39	0.1031	14.9491	39	0.0715	13.2649	39	0.0497	11.8786
40	0.1420	17.1591	40	0.0972	15.0463	40	0.0668	13.3317	40	0.0460	11.9246
41	0.1353	17.2944	41	0.09172	15.1380	41	0.06241	13.3941	41	0.04262	11.9672
42	0.1288	17.4232	42	0.08653	15.2245	42	0.05833	13.4524	42	0.03946	12.0067
43	0.1227	17.5459	43	0.08163	15.3062	43	0.05451	13.5070	43	0.03654	12.0432
44	0.1169	17.6628	44	0.07701	15.3832	44	0.05095	13.5579	44	0.03383	12.0771
45	0.1113	17.7741	45	0.07265	15.4558	45	0.04761	13.6055	45	0.03133	12.1084
46	0.10600	17.8801	46	0.06854	15.5244	46	0.04450	13.6500	46	0.02901	12.1374
47	0.10095	17.9810	47	0.06466	15.5890	47	0.04159	13.6916	47	0.02686	12.1643
48	0.09614	18.0772	48	0.06100	15.6500	48	0.03887	13.7305	48	0.02487	12.1891
49	0.09156	18.1687	49	0.05755	15.7076	49	0.03632	13.7668	49	0.02303	12.2122
50	0.08720	18.2559	50	0.05429	15.7619	50	0.03395	13.8007	50	0.02132	12.2335
55	0.06833	18.6335	55	0.04057	15.9905	55	0.02420	13.9399	55	0.01451	12.3186
60	0.05354	18.9293	60	0.03031	16.1614	60	0.01726	14.0392	60	0.00988	12.3766
65	0.04195	19.1611	65	0.02265	16.2891	65	0.01230	14.1099	65	0.00672	12.4160
70	0.03287	19.3427	70	0.01693	16.3845	70	0.00877	14.1604	70	0.00457	12.4428
75	0.02575	19.4850	75	0.01265	16.4558	75	0.00625	14.1964	75	0.00311	12.4611
80	0.02018	19.5965	80	0.00945	16.5091	80	0.00446	14.2220	80	0.00212	12.4735
85	0.01581	19.6838	85	0.00706	16.5489	85	0.00318	14.2403	85	0.00144	12.4820
90	0.01239	19.7523	90	0.00528	16.5787	90	0.00227	14.2533	90	0.00098	12.4877
95	0.00971	19.8059	95	0.00394	16.6009	95	0.00162	14.2626	95	0.00067	12.4917
100	0.00760	19.8479	100	0.00295	16.6175	100	0.00115	14.2693	100	0.00045	12.4943

259

Discounting tables, 9% to 12%

9%/year rate			10%/year rate			11%/year rate			12%/year rate		
No. of years	PV/FV	PV/RV	No. of years	PV/FV	PV/RV	No. of years	PV/FV	PV/RV	No. of years	PV/FV	PV/RV
1	0.9174	0.9174	1	0.9091	0.9091	1	0.9009	0.9009	1	0.8929	0.8929
2	0.8417	1.7591	2	0.8264	1.7355	2	0.8116	1.7125	2	0.7972	1.6901
3	0.7722	2.5313	3	0.7513	2.4869	3	0.7312	2.4437	3	0.7118	2.4018
4	0.7084	3.2397	4	0.6830	3.1699	4	0.6587	3.1024	4	0.6355	3.0373
5	0.6499	3.8897	5	0.6209	3.7908	5	0.5935	3.6959	5	0.5674	3.6048
6	0.5963	4.4859	6	0.5645	4.3553	6	0.5346	4.2305	6	0.5066	4.1114
7	0.5470	5.0330	7	0.5132	4.8684	7	0.4817	4.7122	7	0.4523	4.5638
8	0.5019	5.5348	8	0.4665	5.3349	8	0.4339	5.1461	8	0.4039	4.9676
9	0.4604	5.9952	9	0.4241	5.7590	9	0.3909	5.5370	9	0.3606	5.3282
10	0.4224	6.4177	10	0.3855	6.1446	10	0.3522	5.8892	10	0.3220	5.6502
11	0.3875	6.8052	11	0.3505	6.4951	11	0.3173	6.2065	11	0.2875	5.9377
12	0.3555	7.1607	12	0.3186	6.8137	12	0.2858	6.4924	12	0.2567	6.1944
13	0.3262	7.4869	13	0.2897	7.1034	13	0.2575	6.7499	13	0.2292	6.4235
14	0.2992	7.7862	14	0.2633	7.3667	14	0.2320	6.9819	14	0.2046	6.6282
15	0.2745	8.0607	15	0.2394	7.6061	15	0.2090	7.1909	15	0.1827	6.8109
16	0.2519	8.3126	16	0.2176	7.8237	16	0.1883	7.3792	16	0.1631	6.9740
17	0.2311	8.5436	17	0.1978	8.0216	17	0.1696	7.5488	17	0.1456	7.1196
18	0.2120	8.7556	18	0.1799	8.2014	18	0.1528	7.7016	18	0.1300	7.2497
19	0.1945	8.9501	19	0.1635	8.3649	19	0.1377	7.8393	19	0.1161	7.3658
20	0.1784	9.1285	20	0.1486	8.5136	20	0.1240	7.9633	20	0.1037	7.4694
21	0.1637	9.2922	21	0.1351	8.6487	21	0.1117	8.0751	21	0.0926	7.5620
22	0.1502	9.4424	22	0.1228	8.7715	22	0.1007	8.1757	22	0.0826	7.6446
23	0.1378	9.5802	23	0.1117	8.8832	23	0.0907	8.2664	23	0.0738	7.7184
24	0.1264	9.7066	24	0.1015	8.9847	24	0.0817	8.3481	24	0.0659	7.7843
25	0.1160	9.8226	25	0.0923	9.0770	25	0.0736	8.4217	25	0.0588	7.8431
26	0.10639	9.9290	26	0.08391	9.1609	26	0.06631	8.4881	26	0.05252	7.8957
27	0.09761	10.0266	27	0.07628	9.2372	27	0.05974	8.5478	27	0.04689	7.9426
28	0.08955	10.1161	28	0.06934	9.3066	28	0.05382	8.6016	28	0.04187	7.9844
29	0.08215	10.1983	29	0.06304	9.3696	29	0.04849	8.6501	29	0.03738	8.0218
30	0.07537	10.2737	30	0.05731	9.4269	30	0.04368	8.6938	30	0.03338	8.0552
31	0.06915	10.3428	31	0.05210	9.4790	31	0.03935	8.7331	31	0.02980	8.0850
32	0.06344	10.4062	32	0.04736	9.5264	32	0.03545	8.7686	32	0.02661	8.1116
33	0.05820	10.4644	33	0.04306	9.5694	33	0.03194	8.8005	33	0.02376	8.1354
34	0.05339	10.5178	34	0.03914	9.6086	34	0.02878	8.8293	34	0.02121	8.1566
35	0.04899	10.5668	35	0.03558	9.6442	35	0.02592	8.8552	35	0.01894	8.1755
36	0.04494	10.6118	36	0.03235	9.6765	36	0.02335	8.8786	36	0.01691	8.1924
37	0.04123	10.6530	37	0.02941	9.7059	37	0.02104	8.8996	37	0.01510	8.2075
38	0.03783	10.6908	38	0.02673	9.7327	38	0.01896	8.9186	38	0.01348	8.2210
39	0.03470	10.7255	39	0.02430	9.7570	39	0.01708	8.9357	39	0.01204	8.2330
40	0.03184	10.7574	40	0.02209	9.7791	40	0.01538	8.9511	40	0.01075	8.2438
41	0.02921	10.7866	41	0.02009	9.7991	41	0.01386	8.9649	41	0.00960	8.2534
42	0.02680	10.8134	42	0.01826	9.8174	42	0.01249	8.9774	42	0.00857	8.2619
43	0.02458	10.8380	43	0.01660	9.8340	43	0.01125	8.9886	43	0.00765	8.2696
44	0.02255	10.8605	44	0.01509	9.8491	44	0.01013	8.9988	44	0.00683	8.2764
45	0.02069	10.8812	45	0.01372	9.8628	45	0.00913	9.0079	45	0.00610	8.2825
46	0.01898	10.9002	46	0.01247	9.8753	46	0.00823	9.0161	46	0.00544	8.2880
47	0.01742	10.9176	47	0.01134	9.8866	47	0.00741	9.0235	47	0.00486	8.2928
48	0.01598	10.9336	48	0.01031	9.8969	48	0.00668	9.0302	48	0.00434	8.2972
49	0.01466	10.9482	49	0.00937	9.9063	49	0.00601	9.0362	49	0.00388	8.3010
50	0.01345	10.9617	50	0.00852	9.9148	50	0.00542	9.0417	50	0.00346	8.3045
55	0.00874	11.0140	55	0.00529	9.9471	55	0.00322	9.0617	55	0.00196	8.3170
60	0.00568	11.0480	60	0.00328	9.9672	60	0.00191	9.0736	60	0.00111	8.3240
65	0.00369	11.0701	65	0.00204	9.9796	65	0.00113	9.0806	65	0.00063	8.3281
70	0.00240	11.0844	70	0.00127	9.9873	70	0.00067	9.0848	70	0.00036	8.3303
75	0.00156	11.0938	75	0.00079	9.9921	75	0.00040	9.0873	75	0.00020	8.3316
80	0.00101	11.0998	80	0.00049	9.9951	80	0.00024	9.0888	80	0.00012	8.3324
85	0.00066	11.1038	85	0.00030	9.9970	85	0.00014	9.0896	85	0.00007	8.3328
90	0.00043	11.1064	90	0.00019	9.9981	90	0.00008	9.0902	90	0.00004	8.3330
95	0.00028	11.1080	95	0.00012	9.9988	95	0.00005	9.0905	95	0.00002	8.3332
100	0.00018	11.1091	100	0.00007	9.9993	100	0.00003	9.0906	100	0.00001	8.3332

Discounting tables, 13% to 16%

13%/year rate			14%/year rate			15%/year rate			16%/year rate		
No. of years	PV/FV	PV/RV	No. of years	PV/FV	PV/RV	No. of years	PV/FV	PV/RV	No. of years	PV/FV	PV/RV
1	0.8850	0.8850	1	0.8772	0.8772	1	0.8696	0.8696	1	0.8621	0.8621
2	0.7831	1.6681	2	0.7695	1.6467	2	0.7561	1.6257	2	0.7432	1.6052
3	0.6931	2.3612	3	0.6750	2.3216	3	0.6575	2.2832	3	0.6407	2.2459
4	0.6133	2.9745	4	0.5921	2.9137	4	0.5718	2.8550	4	0.5523	2.7982
5	0.5428	3.5172	5	0.5194	3.4331	5	0.4972	3.3522	5	0.4761	3.2743
6	0.4803	3.9975	6	0.4556	3.8887	6	0.4323	3.7845	6	0.4104	3.6847
7	0.4251	4.4226	7	0.3996	4.2883	7	0.3759	4.1604	7	0.3538	4.0386
8	0.3762	4.7988	8	0.3506	4.6389	8	0.3269	4.4873	8	0.3050	4.3436
9	0.3329	5.1317	9	0.3075	4.9464	9	0.2843	4.7716	9	0.2630	4.6065
10	0.2946	5.4262	10	0.2697	5.2161	10	0.2472	5.0188	10	0.2267	4.8332
11	0.2607	5.6869	11	0.2366	5.4527	11	0.2149	5.2337	11	0.1954	5.0286
12	0.2307	5.9176	12	0.2076	5.6603	12	0.1869	5.4206	12	0.1685	5.1971
13	0.2042	6.1218	13	0.1821	5.8424	13	0.1625	5.5831	13	0.1452	5.3423
14	0.1807	6.3025	14	0.1597	6.0021	14	0.1413	5.7245	14	0.1252	5.4675
15	0.1599	6.4624	15	0.1401	6.1422	15	0.1229	5.8474	15	0.1079	5.5755
16	0.14150	6.6039	16	0.12289	6.2651	16	0.10686	5.9542	16	0.09304	5.6685
17	0.12522	6.7291	17	0.10780	6.3729	17	0.09293	6.0472	17	0.08021	5.7487
18	0.11081	6.8399	18	0.09456	6.4674	18	0.08081	6.1280	18	0.06914	5.8178
19	0.09806	6.9380	19	0.08295	6.5504	19	0.07027	6.1982	19	0.05961	5.8775
20	0.08678	7.0248	20	0.07276	6.6231	20	0.06110	6.2593	20	0.05139	5.9288
21	0.07680	7.1016	21	0.06383	6.6870	21	0.05313	6.3125	21	0.04430	5.9731
22	0.06796	7.1695	22	0.05599	6.7429	22	0.04620	6.3587	22	0.03819	6.0113
23	0.06014	7.2297	23	0.04911	6.7921	23	0.04017	6.3988	23	0.03292	6.0442
24	0.05323	7.2829	24	0.04308	6.8351	24	0.03493	6.4338	24	0.02838	6.0726
25	0.04710	7.3300	25	0.03779	6.8729	25	0.03038	6.4641	25	0.02447	6.0971
26	0.04168	7.3717	26	0.03315	6.9061	26	0.02642	6.4906	26	0.02109	6.1182
27	0.03689	7.4086	27	0.02908	6.9352	27	0.02297	6.5135	27	0.01818	6.1364
28	0.03264	7.4412	28	0.02551	6.9607	28	0.01997	6.5335	28	0.01567	6.1520
29	0.02889	7.4701	29	0.02237	6.9830	29	0.01737	6.5509	29	0.01351	6.1656
30	0.02557	7.4957	30	0.01963	7.0027	30	0.01510	6.5660	30	0.01165	6.1772
31	0.02262	7.5183	31	0.01722	7.0199	31	0.01313	6.5791	31	0.01004	6.1872
32	0.02002	7.5383	32	0.01510	7.0350	32	0.01142	6.5905	32	0.00866	6.1959
33	0.01772	7.5560	33	0.01325	7.0482	33	0.00993	6.6005	33	0.00746	6.2034
34	0.01568	7.5717	34	0.01162	7.0599	34	0.00864	6.6091	34	0.00643	6.2098
35	0.01388	7.5856	35	0.01019	7.0700	35	0.00751	6.6166	35	0.00555	6.2153
36	0.01228	7.5979	36	0.00894	7.0790	36	0.00653	6.6231	36	0.00478	6.2201
37	0.01087	7.6087	37	0.00784	7.0868	37	0.00568	6.6288	37	0.00412	6.2242
38	0.00962	7.6183	38	0.00688	7.0937	38	0.00494	6.6338	38	0.00355	6.2278
39	0.00851	7.6268	39	0.00604	7.0997	39	0.00429	6.6380	39	0.00306	6.2309
40	0.00753	7.6344	40	0.00529	7.1050	40	0.00373	6.6418	40	0.00264	6.2335
41	0.00666	7.6410	41	0.00464	7.1097	41	0.00325	6.6450	41	0.00228	6.2358
42	0.00590	7.6469	42	0.00407	7.1138	42	0.00282	6.6478	42	0.00196	6.2377
43	0.00522	7.6522	43	0.00357	7.1173	43	0.00245	6.6503	43	0.00169	6.2394
44	0.00462	7.6568	44	0.00313	7.1205	44	0.00213	6.6524	44	0.00146	6.2409
45	0.00409	7.6609	45	0.00275	7.1232	45	0.00186	6.6543	45	0.00126	6.2421
46	0.00362	7.6645	46	0.00241	7.1256	46	0.00161	6.6559	46	0.00108	6.2432
47	0.00320	7.6677	47	0.00212	7.1277	47	0.00140	6.6573	47	0.00093	6.2442
48	0.00283	7.6705	48	0.00186	7.1296	48	0.00122	6.6585	48	0.00081	6.2450
49	0.00251	7.6730	49	0.00163	7.1312	49	0.00106	6.6596	49	0.00069	6.2457
50	0.00222	7.6752	50	0.00143	7.1327	50	0.00092	6.6605	50	0.00060	6.2463
55	0.00120	7.6830	55	0.00074	7.1376	55	0.00046	6.6636	55	0.00028	6.2482
60	0.00065	7.6873	60	0.00039	7.1401	60	0.00023	6.6651	60	0.00014	6.2492
65	0.00035	7.6896	65	0.00020	7.1414	65	0.00011	6.6659	65	0.00006	6.2496
70	0.00019	7.6908	70	0.00010	7.1421	70	0.00006	6.6663	70	0.00003	6.2498
75	0.00010	7.6915	75	0.00005	7.1425	75	0.00003	6.6665	75	0.00001	6.2499
80	0.00006	7.6919	80	0.00003	7.1427	80	0.00001	6.6666	80	0.00001	6.2500
85	0.00003	7.6921	85	0.00001	7.1428	85	0.00001	6.6666	85	0.00000	6.2500
90	0.00002	7.6922	90	0.00001	7.1428	90	0.00000	6.6666	90	0.00000	6.2500
95	0.00001	7.6922	95	0.00000	7.1428	95	0.00000	6.6667	95	0.00000	6.2500
100	0.00000	7.6923	100	0.00000	7.1428	100	0.00000	6.6667	100	0.00000	6.2500

Discounting tables, 17% to 20%

17%/year rate			18%/year rate			19%/year rate			20%/year rate		
No. of years	PV/FV	PV/RV	No. of years	PV/FV	PV/RV	No. of years	PV/FV	PV/RV	No. of years	PV/FV	PV/RV
1	0.8547	0.8547	1	0.8475	0.8475	1	0.8403	0.8403	1	0.8333	0.8333
2	0.7305	1.5852	2	0.7182	1.5656	2	0.7062	1.5465	2	0.6944	1.5278
3	0.6244	2.2096	3	0.6086	2.1743	3	0.5934	2.1399	3	0.5787	2.1065
4	0.5337	2.7432	4	0.5158	2.6901	4	0.4987	2.6386	4	0.4823	2.5887
5	0.4561	3.1993	5	0.4371	3.1272	5	0.4190	3.0576	5	0.4019	2.9906
6	0.3898	3.5892	6	0.3704	3.4976	6	0.3521	3.4098	6	0.3349	3.3255
7	0.3332	3.9224	7	0.3139	3.8115	7	0.2959	3.7057	7	0.2791	3.6046
8	0.2848	4.2072	8	0.2660	4.0776	8	0.2487	3.9544	8	0.2326	3.8372
9	0.2434	4.4506	9	0.2255	4.3030	9	0.2090	4.1633	9	0.1938	4.0310
10	0.2080	4.6586	10	0.1911	4.4941	10	0.1756	4.3389	10	0.1615	4.1925
11	0.1778	4.8364	11	0.1619	4.6560	11	0.1476	4.4865	11	0.1346	4.3271
12	0.1520	4.9884	12	0.1372	4.7932	12	0.1240	4.6105	12	0.1122	4.4392
13	0.1299	5.1183	13	0.1163	4.9095	13	0.1042	4.7147	13	0.0935	4.5327
14	0.1110	5.2293	14	0.0985	5.0081	14	0.0876	4.8023	14	0.0779	4.6106
15	0.0949	5.3242	15	0.0835	5.0916	15	0.0736	4.8759	15	0.0649	4.6755
16	0.08110	5.4053	16	0.07078	5.1624	16	0.06184	4.9377	16	0.05409	4.7296
17	0.06932	5.4746	17	0.05998	5.2223	17	0.05196	4.9897	17	0.04507	4.7746
18	0.05925	5.5339	18	0.05083	5.2732	18	0.04367	5.0333	18	0.03756	4.8122
19	0.05064	5.5845	19	0.04308	5.3162	19	0.03670	5.0700	19	0.03130	4.8435
20	0.04328	5.6278	20	0.03651	5.3527	20	0.03084	5.1009	20	0.02608	4.8696
21	0.03699	5.6648	21	0.03094	5.3837	21	0.02591	5.1268	21	0.02174	4.8913
22	0.03162	5.6964	22	0.02622	5.4099	22	0.02178	5.1486	22	0.01811	4.9094
23	0.02702	5.7234	23	0.02222	5.4321	23	0.01830	5.1668	23	0.01509	4.9245
24	0.02310	5.7465	24	0.01883	5.4509	24	0.01538	5.1822	24	0.01258	4.9371
25	0.01974	5.7662	25	0.01596	5.4669	25	0.01292	5.1951	25	0.01048	4.9476
26	0.01687	5.7831	26	0.01352	5.4804	26	0.01086	5.2060	26	0.00874	4.9563
27	0.01442	5.7975	27	0.01146	5.4919	27	0.00912	5.2151	27	0.00728	4.9636
28	0.01233	5.8099	28	0.00971	5.5016	28	0.00767	5.2228	28	0.00607	4.9697
29	0.01053	5.8204	29	0.00823	5.5098	29	0.00644	5.2292	29	0.00506	4.9747
30	0.00900	5.8294	30	0.00697	5.5168	30	0.00541	5.2347	30	0.00421	4.9789
31	0.00770	5.8371	31	0.00591	5.5227	31	0.00455	5.2392	31	0.00351	4.9824
32	0.00658	5.8437	32	0.00501	5.5277	32	0.00382	5.2430	32	0.00293	4.9854
33	0.00562	5.8493	33	0.00425	5.5320	33	0.00321	5.2462	33	0.00244	4.9878
34	0.00480	5.8541	34	0.00360	5.5356	34	0.00270	5.2489	34	0.00203	4.9898
35	0.00411	5.8582	35	0.00305	5.5386	35	0.00227	5.2512	35	0.00169	4.9915
36	0.00351	5.8617	36	0.00258	5.5412	36	0.00191	5.2531	36	0.00141	4.9929
37	0.00300	5.8647	37	0.00219	5.5434	37	0.00160	5.2547	37	0.00118	4.9941
38	0.00256	5.8673	38	0.00186	5.5452	38	0.00135	5.2561	38	0.00098	4.9951
39	0.00219	5.8695	39	0.00157	5.5468	39	0.00113	5.2572	39	0.00082	4.9959
40	0.00187	5.8713	40	0.00133	5.5482	40	0.00095	5.2582	40	0.00068	4.9966
41	0.00160	5.8729	41	0.00113	5.5493	41	0.00080	5.2590	41	0.00057	4.9972
42	0.00137	5.8743	42	0.00096	5.5502	42	0.00067	5.2596	42	0.00047	4.9976
43	0.00117	5.8755	43	0.00081	5.5510	43	0.00056	5.2602	43	0.00039	4.9980
44	0.00100	5.8765	44	0.00069	5.5517	44	0.00047	5.2607	44	0.00033	4.9984
45	0.00085	5.8773	45	0.00058	5.5523	45	0.00040	5.2611	45	0.00027	4.9986
46	0.00073	5.8781	46	0.00049	5.5528	46	0.00033	5.2614	46	0.00023	4.9989
47	0.00062	5.8787	47	0.00042	5.5532	47	0.00028	5.2617	47	0.00019	4.9991
48	0.00053	5.8792	48	0.00035	5.5536	48	0.00024	5.2619	48	0.00016	4.9992
49	0.00046	5.8797	49	0.00030	5.5539	49	0.00020	5.2621	49	0.00013	4.9993
50	0.00039	5.8801	50	0.00025	5.5541	50	0.00017	5.2623	50	0.00011	4.9995
55	0.00018	5.8813	55	0.00011	5.5549	55	0.00007	5.2628	55	0.00004	4.9998
60	0.00008	5.8819	60	0.00005	5.5553	60	0.00003	5.2630	60	0.00002	4.9999
65	0.00004	5.8821	65	0.00002	5.5554	65	0.00001	5.2631	65	0.00001	5.0000
70	0.00002	5.8823	70	0.00001	5.5555	70	0.00001	5.2631	70	0.00000	5.0000
75	0.00001	5.8823	75	0.00000	5.5555	75	0.00000	5.2631	75	0.00000	5.0000
100	0.00000	5.8824	100	0.00000	5.5556	100	0.00000	5.2632	100	0.00000	5.0000

Discounting tables, 22% to 28%

22%/year rate			24%/year rate			26%/year rate			28%/year rate		
No. of years	PV/FV	PV/RV	No. of years	PV/FV	PV/RV	No. of years	PV/FV	PV/RV	No. of years	PV/FV	PV/RV
1	0.8197	0.8197	1	0.8065	0.8065	1	0.7937	0.7937	1	0.7813	0.7813
2	0.6719	1.4915	2	0.6504	1.4568	2	0.6299	1.4235	2	0.6104	1.3916
3	0.5507	2.0422	3	0.5245	1.9813	3	0.4999	1.9234	3	0.4768	1.8684
4	0.4514	2.4936	4	0.4230	2.4043	4	0.3968	2.3202	4	0.3725	2.2410
5	0.3700	2.8636	5	0.3411	2.7454	5	0.3149	2.6351	5	0.2910	2.5320
6	0.3033	3.1669	6	0.2751	3.0205	6	0.2499	2.8850	6	0.2274	2.7594
7	0.2486	3.4155	7	0.2218	3.2423	7	0.1983	3.0833	7	0.1776	2.9370
8	0.2038	3.6193	8	0.1789	3.4212	8	0.1574	3.2407	8	0.1388	3.0758
9	0.1670	3.7863	9	0.1443	3.5655	9	0.1249	3.3657	9	0.1084	3.1842
10	0.1369	3.9232	10	0.1164	3.6819	10	0.0992	3.4648	10	0.0847	3.2689
11	0.11221	4.0354	11	0.09383	3.7757	11	0.07869	3.5435	11	0.06617	3.3351
12	0.09198	4.1274	12	0.07567	3.8514	12	0.06245	3.6059	12	0.05170	3.3868
13	0.07539	4.2028	13	0.06103	3.9124	13	0.04957	3.6555	13	0.04039	3.4272
14	0.06180	4.2646	14	0.04921	3.9616	14	0.03934	3.6949	14	0.03155	3.4587
15	0.05065	4.3152	15	0.03969	4.0013	15	0.03122	3.7261	15	0.02465	3.4834
16	0.04152	4.3567	16	0.03201	4.0333	16	0.02478	3.7509	16	0.01926	3.5026
17	0.03403	4.3908	17	0.02581	4.0591	17	0.01967	3.7705	17	0.01505	3.5177
18	0.02789	4.4187	18	0.02082	4.0799	18	0.01561	3.7861	18	0.01175	3.5294
19	0.02286	4.4415	19	0.01679	4.0967	19	0.01239	3.7985	19	0.00918	3.5386
20	0.01874	4.4603	20	0.01354	4.1103	20	0.00983	3.8083	20	0.00717	3.5458
21	0.01536	4.4756	21	0.01092	4.1212	21	0.00780	3.8161	21	0.00561	3.5514
22	0.01259	4.4882	22	0.00880	4.1300	22	0.00619	3.8223	22	0.00438	3.5558
23	0.01032	4.4985	23	0.00710	4.1371	23	0.00491	3.8273	23	0.00342	3.5592
24	0.00846	4.5070	24	0.00573	4.1428	24	0.00390	3.8312	24	0.00267	3.5619
25	0.00693	4.5139	25	0.00462	4.1474	25	0.00310	3.8342	25	0.00209	3.5640
26	0.00568	4.5196	26	0.00372	4.1511	26	0.00246	3.8367	26	0.00163	3.5656
27	0.00466	4.5243	27	0.00300	4.1542	27	0.00195	3.8387	27	0.00127	3.5669
28	0.00382	4.5281	28	0.00242	4.1566	28	0.00155	3.8402	28	0.00100	3.5679
29	0.00313	4.5312	29	0.00195	4.1585	29	0.00123	3.8414	29	0.00078	3.5687
30	0.00257	4.5338	30	0.00158	4.1601	30	0.00097	3.8424	30	0.00061	3.5693
31	0.00210	4.5359	31	0.00127	4.1614	31	0.00077	3.8432	31	0.00047	3.5697
32	0.00172	4.5376	32	0.00102	4.1624	32	0.00061	3.8438	32	0.00037	3.5701
33	0.00141	4.5390	33	0.00083	4.1632	33	0.00049	3.8443	33	0.00029	3.5704
34	0.00116	4.5402	34	0.00067	4.1639	34	0.00039	3.8447	34	0.00023	3.5706
35	0.00095	4.5411	35	0.00054	4.1644	35	0.00031	3.8450	35	0.00018	3.5708
36	0.00078	4.5419	36	0.00043	4.1649	36	0.00024	3.8452	36	0.00014	3.5709
37	0.00064	4.5426	37	0.00035	4.1652	37	0.00019	3.8454	37	0.00011	3.5710
38	0.00052	4.5431	38	0.00028	4.1655	38	0.00015	3.8456	38	0.00008	3.5711
39	0.00043	4.5435	39	0.00023	4.1657	39	0.00012	3.8457	39	0.00007	3.5712
40	0.00035	4.5439	40	0.00018	4.1659	40	0.00010	3.8458	40	0.00005	3.5712
41	0.00029	4.5441	41	0.00015	4.1661	41	0.00008	3.8459	41	0.00004	3.5713
42	0.00024	4.5444	42	0.00012	4.1662	42	0.00006	3.8459	42	0.00003	3.5713
43	0.00019	4.5446	43	0.00010	4.1663	43	0.00005	3.8460	43	0.00002	3.5713
44	0.00016	4.5447	44	0.00008	4.1663	44	0.00004	3.8460	44	0.00002	3.5714
45	0.00013	4.5449	45	0.00006	4.1664	45	0.00003	3.8460	45	0.00001	3.5714
46	0.00011	4.5450	46	0.00005	4.1665	46	0.00002	3.8461	46	0.00001	3.5714
47	0.00009	4.5451	47	0.00004	4.1665	47	0.00002	3.8461	47	0.00001	3.5714
48	0.00007	4.5451	48	0.00003	4.1665	48	0.00002	3.8461	48	0.00001	3.5714
49	0.00006	4.5452	49	0.00003	4.1666	49	0.00001	3.8461	49	0.00001	3.5714
50	0.00005	4.5452	50	0.00002	4.1666	50	0.00001	3.8461	50	0.00000	3.5714
55	0.00002	4.5454	55	0.00001	4.1666	55	0.00000	3.8461	55	0.00000	3.5714
60	0.00001	4.5454	60	0.00000	4.1667	60	0.00000	3.8462	60	0.00000	3.5714
65	0.00000	4.5454	65	0.00000	4.1667	65	0.00000	3.8462	65	0.00000	3.5714
100	0.00000	4.5455	100	0.00000	4.1667	100	0.00000	3.8462	100	0.00000	3.5714

Discounting tables, 30% to 36%

30%/year rate			32%/year rate			34%/year rate			36%/year rate		
No. of years	PV/FV	PV/RV	No. of years	PV/FV	PV/RV	No. of years	PV/FV	PV/RV	No. of years	PV/FV	PV/RV
1	0.7692	0.7692	1	0.7576	0.7576	1	0.7463	0.7463	1	0.7353	0.7353
2	0.5917	1.3609	2	0.5739	1.3315	2	0.5569	1.3032	2	0.5407	1.2760
3	0.4552	1.8161	3	0.4348	1.7663	3	0.4156	1.7188	3	0.3975	1.6735
4	0.3501	2.1662	4	0.3294	2.0957	4	0.3102	2.0290	4	0.2923	1.9658
5	0.2693	2.4356	5	0.2495	2.3452	5	0.2315	2.2604	5	0.2149	2.1807
6	0.20718	2.6427	6	0.18904	2.5342	6	0.17273	2.4331	6	0.15804	2.3388
7	0.15937	2.8021	7	0.14321	2.6775	7	0.12890	2.5620	7	0.11621	2.4550
8	0.12259	2.9247	8	0.10849	2.7860	8	0.09620	2.6582	8	0.08545	2.5404
9	0.09430	3.0190	9	0.08219	2.8681	9	0.07179	2.7300	9	0.06283	2.6033
10	0.07254	3.0915	10	0.06227	2.9304	10	0.05357	2.7836	10	0.04620	2.6495
11	0.05580	3.1473	11	0.04717	2.9776	11	0.03998	2.8236	11	0.03397	2.6834
12	0.04292	3.1903	12	0.03574	3.0133	12	0.02984	2.8534	12	0.02498	2.7084
13	0.03302	3.2233	13	0.02707	3.0404	13	0.02227	2.8757	13	0.01837	2.7268
14	0.02540	3.2487	14	0.02051	3.0609	14	0.01662	2.8923	14	0.01350	2.7403
15	0.01954	3.2682	15	0.01554	3.0764	15	0.01240	2.9047	15	0.00993	2.7502
16	0.01503	3.2832	16	0.01177	3.0882	16	0.00925	2.9140	16	0.00730	2.7575
17	0.01156	3.2948	17	0.00892	3.0971	17	0.00691	2.9209	17	0.00537	2.7629
18	0.00889	3.3037	18	0.00676	3.1039	18	0.00515	2.9260	18	0.00395	2.7668
19	0.00684	3.3105	19	0.00512	3.1090	19	0.00385	2.9299	19	0.00290	2.7697
20	0.00526	3.3158	20	0.00388	3.1129	20	0.00287	2.9327	20	0.00213	2.7718
21	0.00405	3.3198	21	0.00294	3.1158	21	0.00214	2.9349	21	0.00157	2.7734
22	0.00311	3.3230	22	0.00223	3.1180	22	0.00160	2.9365	22	0.00115	2.7746
23	0.00239	3.3254	23	0.00169	3.1197	23	0.00119	2.9377	23	0.00085	2.7754
24	0.00184	3.3272	24	0.00128	3.1210	24	0.00089	2.9386	24	0.00062	2.7760
25	0.00142	3.3286	25	0.00097	3.1220	25	0.00066	2.9392	25	0.00046	2.7765
26	0.00109	3.3297	26	0.00073	3.1227	26	0.00050	2.9397	26	0.00034	2.7768
27	0.00084	3.3305	27	0.00056	3.1233	27	0.00037	2.9401	27	0.00025	2.7771
28	0.00065	3.3312	28	0.00042	3.1237	28	0.00028	2.9404	28	0.00018	2.7773
29	0.00050	3.3317	29	0.00032	3.1240	29	0.00021	2.9406	29	0.00013	2.7774
30	0.00038	3.3321	30	0.00024	3.1242	30	0.00015	2.9407	30	0.00010	2.7775
31	0.00029	3.3324	31	0.00018	3.1244	31	0.00011	2.9408	31	0.00007	2.7776
32	0.00023	3.3326	32	0.00014	3.1246	32	0.00009	2.9409	32	0.00005	2.7776
33	0.00017	3.3328	33	0.00010	3.1247	33	0.00006	2.9410	33	0.00004	2.7777
34	0.00013	3.3329	34	0.00008	3.1248	34	0.00005	2.9410	34	0.00003	2.7777
35	0.00010	3.3330	35	0.00006	3.1248	35	0.00004	2.9411	35	0.00002	2.7777
36	0.00008	3.3331	36	0.00005	3.1249	36	0.00003	2.9411	36	0.00002	2.7777
37	0.00006	3.3331	37	0.00003	3.1249	37	0.00002	2.9411	37	0.00001	2.7777
38	0.00005	3.3332	38	0.00003	3.1249	38	0.00001	2.9411	38	0.00001	2.7778
39	0.00004	3.3332	39	0.00002	3.1249	39	0.00001	2.9411	39	0.00001	2.7778
40	0.00003	3.3332	40	0.00002	3.1250	40	0.00001	2.9412	40	0.00000	2.7778
41	0.00002	3.3333	41	0.00001	3.1250	41	0.00001	2.9412	41	0.00000	2.7778
42	0.00002	3.3333	42	0.00001	3.1250	42	0.00000	2.9412	42	0.00000	2.7778
43	0.00001	3.3333	43	0.00001	3.1250	43	0.00000	2.9412	43	0.00000	2.7778
44	0.00001	3.3333	44	0.00000	3.1250	44	0.00000	2.9412	44	0.00000	2.7778
45	0.00001	3.3333	45	0.00000	3.1250	45	0.00000	2.9412	45	0.00000	2.7778
46	0.00001	3.3333	46	0.00000	3.1250	46	0.00000	2.9412	46	0.00000	2.7778
47	0.00000	3.3333	47	0.00000	3.1250	47	0.00000	2.9412	47	0.00000	2.7778
48	0.00000	3.3333	48	0.00000	3.1250	48	0.00000	2.9412	48	0.00000	2.7778
49	0.00000	3.3333	49	0.00000	3.1250	49	0.00000	2.9412	49	0.00000	2.7778
50	0.00000	3.3333	50	0.00000	3.1250	50	0.00000	2.9412	50	0.00000	2.7778
100	0.00000	3.3333	100	0.00000	3.1250	100	0.00000	2.9412	100	0.00000	2.7778

Discounting tables, 40% to 55%

40%/year rate			45%/year rate			50%/year rate			55%/year rate		
No. of years	PV/FV	PV/RV	No. of years	PV/FV	PV/RV	No. of years	PV/FV	PV/RV	No. of years	PV/FV	PV/RV
1	0.7143	0.7143	1	0.6897	0.6897	1	0.6667	0.6667	1	0.6452	0.6452
2	0.5102	1.2245	2	0.4756	1.1653	2	0.4444	1.1111	2	0.4162	1.0614
3	0.3644	1.5889	3	0.3280	1.4933	3	0.2963	1.4074	3	0.2685	1.3299
4	0.2603	1.8492	4	0.2262	1.7195	4	0.1975	1.6049	4	0.1732	1.5032
5	0.1859	2.0352	5	0.1560	1.8755	5	0.1317	1.7366	5	0.1118	1.6150
6	0.13281	2.1680	6	0.10759	1.9831	6	0.08779	1.8244	6	0.07211	1.6871
7	0.09486	2.2628	7	0.07420	2.0573	7	0.05853	1.8829	7	0.04652	1.7336
8	0.06776	2.3306	8	0.05117	2.1085	8	0.03902	1.9220	8	0.03002	1.7636
9	0.04840	2.3790	9	0.03529	2.1438	9	0.02601	1.9480	9	0.01936	1.7830
10	0.03457	2.4136	10	0.02434	2.1681	10	0.01734	1.9653	10	0.01249	1.7955
11	0.02469	2.4383	11	0.01679	2.1849	11	0.01156	1.9769	11	0.00806	1.8035
12	0.01764	2.4559	12	0.01158	2.1965	12	0.00771	1.9846	12	0.00520	1.8087
13	0.01260	2.4685	13	0.00798	2.2045	13	0.00514	1.9897	13	0.00335	1.8121
14	0.00900	2.4775	14	0.00551	2.2100	14	0.00343	1.9931	14	0.00216	1.8142
15	0.00643	2.4839	15	0.00380	2.2138	15	0.00228	1.9954	15	0.00140	1.8156
16	0.00459	2.4885	16	0.00262	2.2164	16	0.00152	1.9970	16	0.00090	1.8165
17	0.00328	2.4918	17	0.00181	2.2182	17	0.00101	1.9980	17	0.00058	1.8171
18	0.00234	2.4941	18	0.00125	2.2195	18	0.00068	1.9986	18	0.00037	1.8175
19	0.00167	2.4958	19	0.00086	2.2203	19	0.00045	1.9991	19	0.00024	1.8177
20	0.00120	2.4970	20	0.00059	2.2209	20	0.00030	1.9994	20	0.00016	1.8179
21	0.00085	2.4979	21	0.00041	2.2213	21	0.00020	1.9996	21	0.00010	1.8180
22	0.00061	2.4985	22	0.00028	2.2216	22	0.00013	1.9997	22	0.00006	1.8181
23	0.00044	2.4989	23	0.00019	2.2218	23	0.00009	1.9998	23	0.00004	1.8181
24	0.00031	2.4992	24	0.00013	2.2219	24	0.00006	1.9999	24	0.00003	1.8181
25	0.00022	2.4994	25	0.00009	2.2220	25	0.00004	1.9999	25	0.00002	1.8182
26	0.00016	2.4996	26	0.00006	2.2221	26	0.00003	1.9999	26	0.00001	1.8182
27	0.00011	2.4997	27	0.00004	2.2221	27	0.00002	2.0000	27	0.00001	1.8182
28	0.00008	2.4998	28	0.00003	2.2222	28	0.00001	2.0000	28	0.00000	1.8182
29	0.00006	2.4999	29	0.00002	2.2222	29	0.00001	2.0000	29	0.00000	1.8182
30	0.00004	2.4999	30	0.00001	2.2222	30	0.00001	2.0000	30	0.00000	1.8182
31	0.00003	2.4999	31	0.00001	2.2222	31	0.00000	2.0000	31	0.00000	1.8182
32	0.00002	2.4999	32	0.00001	2.2222	32	0.00000	2.0000	32	0.00000	1.8182
33	0.00002	2.5000	33	0.00000	2.2222	33	0.00000	2.0000	33	0.00000	1.8182
34	0.00001	2.5000	34	0.00000	2.2222	34	0.00000	2.0000	34	0.00000	1.8182
35	0.00001	2.5000	35	0.00000	2.2222	35	0.00000	2.0000	35	0.00000	1.8182
100	0.00000	2.5000	100	0.00000	2.2222	100	0.00000	2.0000	100	0.00000	1.8182

Discounting tables, 60% to 75%

60%/year rate			65%/year rate			70%/year rate			75%/year rate		
No. of years	PV/FV	PV/RV	No. of years	PV/FV	PV/RV	No. of years	PV/FV	PV/RV	No. of years	PV/FV	PV/RV
1	0.62500	0.6250	1	0.60606	0.6061	1	0.58824	0.5882	1	0.57143	0.5714
2	0.39063	1.0156	2	0.36731	0.9734	2	0.34602	0.9343	2	0.32653	0.8980
3	0.24414	1.2598	3	0.22261	1.1960	3	0.20354	1.1378	3	0.18659	1.0845
4	0.15259	1.4124	4	0.13492	1.3309	4	0.11973	1.2575	4	0.10662	1.1912
5	0.09537	1.5077	5	0.08177	1.4127	5	0.07043	1.3280	5	0.06093	1.2521
6	0.05960	1.5673	6	0.04956	1.4622	6	0.04143	1.3694	6	0.03482	1.2869
7	0.03725	1.6046	7	0.03003	1.4923	7	0.02437	1.3938	7	0.01989	1.3068
8	0.02328	1.6279	8	0.01820	1.5105	8	0.01434	1.4081	8	0.01137	1.3182
9	0.01455	1.6424	9	0.01103	1.5215	9	0.00843	1.4165	9	0.00650	1.3247
10	0.00909	1.6515	10	0.00669	1.5282	10	0.00496	1.4215	10	0.00371	1.3284
11	0.00568	1.6572	11	0.00405	1.5322	11	0.00292	1.4244	11	0.00212	1.3305
12	0.00355	1.6607	12	0.00246	1.5347	12	0.00172	1.4261	12	0.00121	1.3317
13	0.00222	1.6630	13	0.00149	1.5362	13	0.00101	1.4271	13	0.00069	1.3324
14	0.00139	1.6644	14	0.00090	1.5371	14	0.00059	1.4277	14	0.00040	1.3328
15	0.00087	1.6652	15	0.00055	1.5376	15	0.00035	1.4281	15	0.00023	1.3330
16	0.00054	1.6658	16	0.00033	1.5380	16	0.00021	1.4283	16	0.00013	1.3332
17	0.00034	1.6661	17	0.00020	1.5382	17	0.00012	1.4284	17	0.00007	1.3332
18	0.00021	1.6663	18	0.00012	1.5383	18	0.00007	1.4285	18	0.00004	1.3333
19	0.00013	1.6664	19	0.00007	1.5383	19	0.00004	1.4285	19	0.00002	1.3333
20	0.00008	1.6665	20	0.00004	1.5384	20	0.00002	1.4285	20	0.00001	1.3333
21	0.00005	1.6666	21	0.00003	1.5384	21	0.00001	1.4286	21	0.00001	1.3333
22	0.00003	1.6666	22	0.00002	1.5384	22	0.00001	1.4286	22	0.00000	1.3333
23	0.00002	1.6666	23	0.00001	1.5384	23	0.00001	1.4286	23	0.00000	1.3333
24	0.00001	1.6666	24	0.00001	1.5385	24	0.00000	1.4286	24	0.00000	1.3333
25	0.00001	1.6667	25	0.00000	1.5385	25	0.00000	1.4286	25	0.00000	1.3333
100	0.00000	1.6667	100	0.00000	1.5385	100	0.00000	1.4286	100	0.00000	1.3333

K

Bibliography

This appendix lists some works out of the extensive published literature, including all works referred to in the text. Readers who wish to find up-to-date papers and book reviews should look at one or more of the journals listed in a separate section at the end of this appendix, and/ or on the internet using search words from this book.

Abelson, P. (1996). *Project Appraisal and Evaluation of the Environment*. Macmillan, London. ISBN 0-333-63916-2 (UK) and 0-312-12984-X (USA).
  Aspects covered in case studies are water quality, electricity, slum improvement, mass transit, forests, and soils.
Adler, H.A. (1987). *Economic Appraisal of Transport Projects*. Johns Hopkins University Press, Baltimore, OH (for the World Bank). ISBN 0-8018-3429-5.
  A practical manual on CBA for transport projects, covering developed and developing countries.
Adler, M.D. and Posner, E.A. (2006). *New Foundations of Cost–Benefit Analysis*. Harvard University Press, Harvard, CT. ISBN 0-674-02279-3.
  An American academic critique of CBA as a decision tool.
Arrow, K. *et al.* (1993). See NOAA (1993).
  Also sometimes referred to as Arrow, K., Solow, R. *et al.*
Asian Development Bank (1997 and updates).
  The ADB issues guidelines and updates them from time to time; they contain general descriptions of CBA techniques as well as statements of the conventions required by this particular financing agency. They are often made available to analysts working on ADB-financed projects. This book refers to the *Guidelines for the Economic Analysis of Projects*, dated January 1997 and still in use in 2010.
Asian Development Bank (2002 and updates).
  The ADB also issues handbooks and technical notes, available on the internet. Of particular interest for uncertainty are: *ERD Technical Note No. 2, Integrating Risk into ADB's Economic Analysis of Projects* (2002) and *Handbook for Integrating Risk Analysis in the Economic Analysis of Projects* (2002).

Balcombe, K.G. and Smith, L.E.D. (1999). Refining the use of Monte-Carlo techniques for risk analysis in project planning. *Journal of Development Studies* **36**(2), 113–135.

Discusses ways of including trends, cycles and correlations. (Reprinted in Harberger and Jenkins, 2002.)

Bateman, I., Turner, R.K. and Bateman, S. (1993). Extending cost–benefit analysis of UK highway proposals: environmental evaluation and equity. *Project Appraisal* **8**(4), 213–224.

Discusses COBA, its merits, limitations, and possible improvements.

Belli, P., Anderson, J.R., Barnum, H.N., Dixon, J.A. and Tan, J.-P. (2001). *Economic Analysis of Investment Operations; Analytical Tools and Practical Applications*. World Bank Institute, Washington, DC. ISBN 0-8213-4850-7.

A slightly modified version of the earlier World Bank handbook (World Bank, 1996), and a useful source of authoritative guidance and commentary on overall design and evaluation of development projects. Financial analysis often concerns a project implementing agency. Emphasis on gainers and losers. Many examples.

Belton, V. (1989). The use of a simple multi-criteria model to assist in selection from a shortlist. In: *Readings in Decision Analysis* (ed. S. French). Chapman & Hall, London.

Hierarchy of objectives.

Belton, V. and Stewart, T.J. (2002). *Multiple Criteria Decision Analysis – An Integrated Approach*. Kluwer Academic, Boston, MA. ISBN 0-7923-7505-X.

A thorough description and explanation of many decision aids by two leading academics in the field.

Bowers, J. and Hopkinson, P. (1994). The treatment of landscape in project appraisal: consumption and sustainable approaches. *Project Appraisal* **9**(2), 110–118.

Brealey, R.A. and Myers, S.C. (1984). *Principles of Corporate Finance*, 2nd edn. McGraw-Hill, New York.

Brent, R.J. (1990). *Project Appraisal for Developing Countries*. Harvester Wheat-sheaf, New York. ISBN 0-7450-0422-9.

See Brent (1998).

Brent, R.J. (1998). *Cost–Benefit Analysis for Developing Countries*. Edward Elgar, Cheltenham. ISBN 1-84064-442-7.

This and the similar 1990 book are based on a course at Fordham University. They provide a thorough exposition of the Squire and van der Tak methodology (using the foreign exchange numéraire), and also discuss different methods and opinions to some extent; intended for economists.

Brent, R.J. (1994). Counting and double-counting in project appraisal. *Project Appraisal* **9**(4), 275–281.

Brent, R.J. (2003). *Cost–Benefit Analysis and Health Care Evaluation*. Edward Elgar, Cheltenham. ISBN 1-84064-844-9.

Written by an academic economist, but stresses that non-economists can evaluate healthcare interventions.

Brent, R.J. (2006). *Applied Cost–Benefit Analysis*, 2nd edn. Edward Elgar, Cheltenham. ISBN 1-84376-891-7.

A thorough academic economist-styled textbook, including many application examples. No discussion of choice of numéraire.

Brent, R.J. (ed.) (2009). *Handbook of Research on Cost–Benefit Analysis*. Edward Elgar, Cheltenham. ISBN 9781-84720-069-3.

Collection of papers by 26 contributors, including specialist papers on CBA for health, education, transport and environmental projects.

Carroll, L. (1872). *Through the Looking Glass*. Macmillan, London.

Chervel, M. (1992). La methode des effets trente ans apres. *Tiers-Monde* **33**(132), 873–892.

Chervel, M. and Le Gall, C. (1976). *Manuel d'Évaluation Économique des Projets; la Méthode des Effets*. Ministère de la Coopération, Paris.

Chervel, M. and Le Gall, C. (1978). *Manual of Economic Evaluation of Projects; the Effects Method*. Ministère de la Coopération, Paris.

English translation of the 1976 manual; difficult to obtain.

Cline, W.C. (1993). Give greenhouse abatement a fair chance. *Finance and Development* **30**(1), 3–5.

Part of a special issue on discount rates.

COBA (1996 and later updates). Economic assessment of road schemes. In: *Design Manual for Roads and Bridges*, sect. 1, vol. 13. Her Majesty's Stationery Office, London. ISBN 0115511806.

The user guide for the UK government's computer cost–benefit analysis program (COBA) for roads; successor to several earlier versions. Ringbinder format for periodic piecemeal updating, with CD ROM, or latest version always available on-line at http://www.dft.gov.uk/pgr/economics/software.

Cohn, E. and Johnes, G. (eds) (1994). *Recent Developments in the Economics of Education*. Edward Elgar, Aldershot. ISBN 1-85278-828-3.

Coker, A. and Richards, C. (eds) (1992). *Valuing the Environment – Economic Approaches to Environmental Evaluation*. Belhaven Press (John Wiley, Chichester, 1995). ISBN 1-85293-212-0.

A collection of papers from a 1990 workshop: fundamental discussion of valuation, plus UK case studies.

A large collection of papers, including many on costs and policies.

Culyer, A.J. (ed.) (1991). *The Economics of Health*, vols I and II. Edward Elgar, Aldershot. ISBN 1-85278-176-9.

A collection of papers on many aspects of health economics.

Curry, S. and Weiss, J. (1993). *Project Analysis in Developing Countries*. Macmillan, Basingstoke. ISBN 0-312-09432-9.

A thorough textbook (300 pages) on economic analysis for developing countries (see Curry and Weiss, 2000).

Curry, S. and Weiss, J. (2000). *Cost–Benefit Analysis in Developing Countries*. Palgrave, Basingstoke. ISBN 0333792920.

Extended second edition of the 1993 book, good on economic/financial distinction, and on both numéraires.

Curtin, T.R.C. (1996). Project appraisal and human capital theory. *Project Appraisal* **11**(2), 66–78.

DCLG (2009). *Multi-criteria Analysis: A Manual.* Department for Communities and Local Government, London. Available from: http://www.communities. gov.uk. ISBN 978-1-4098-1023-0.

The British government's official guide on MCA, resulting from experience in the use of HMSO (1991) in intervening years. A thorough description of MCA techniques.

Dixon, J.A., Scura, L.F., Carpenter, R.A. and Sherman, P.B. (1994). *Economic Analysis of Environmental Projects.* Earthscan Publications/Asian Development Bank/World Bank, London. ISBN 1-85383-185-9.

A systematic account of valuation methods for environmental impacts, with extensive case studies.

Eckstein, S. and Lecker, T. (1995). The diminishing marginal effectiveness of the discount rate as a screening device. *Project Appraisal* **10**(1), 49–55.

Examines some unusually shaped financial cash flows.

Environment Agency (2010). *Flood and Coastal Erosion Risk Management Appraisal Guidance* (FCERM-AG). Available at: http://www.environment-agency.gov.uk/research/planning/116705.aspx.

British government's 'living draft', for comment and amendment; perhaps later for publication.

Fabré, P. and Yung, J.M. (1985). *Manuel d'Évaluation des Projets de Développement Rural.* Ministère des Relations Extérieures – Coopération et Développement, Paris.

FAO (1986). *Guide for Training in the Formulation of Agricultural and Rural Investment Projects, Phase 4, Analysis of Expected Results.* Food and Agriculture Organisation, United Nations, Rome.

Field, B.C. (1994). *Environmental Economics – An Introduction.* McGraw-Hill, New York. ISBN 0-07-020797-6.

Aimed at economics and other students with little prior knowledge of economics, generally in a USA context; it includes a basic explanation of willingness-to-pay and opportunity cost (Chapter 3), as well as environmental issues.

Finney, C.E. (1990). A consultant's criteria for the economic ranking of public sector projects. *Project Appraisal* March, 19–22.

Finney, C.E. (1996). The benefits of land levelling on irrigation schemes in Turkey and Sindh province, Pakistan. *Journal of International Commission on Irrigation and Drainage* **45**(1), 21–37.

Fouirry, J.-P. (1996). Cost–benefit analysis in the project cycle, the example of the French-speaking African countries. In: *Cost–Benefit Analysis and Project Appraisal in Developing Countries* (C. Kirkpatrick and J. Weiss, eds). Edward Elgar, Aldershot.

Franck, B. (1996). The effects method and economic cost–benefit analysis: substitutes or complements? In *Cost–Benefit Analysis and Project Appraisal in Developing Countries* (C. Kirkpatrick and J. Weiss, eds). Edward Elgar, Aldershot.

French, S. (ed.) (1989). *Readings in Decision Analysis*. Chapman & Hall, London. ISBN 0-412-32170-X.

A collection of papers from publications of the Operational Research Society of Great Britain, with notes and explanations by French. Thorough and fundamental, gives many references.

Gittinger, J.P. (1982) *Economic Analysis of Agricultural Projects*, 2nd edn. Johns Hopkins University Press, Baltimore, OH (for the World Bank). ISBN 0-8018-2913-5.

For non-economists this book has the advantage of using relatively little economics jargon. Good index combined with glossary.

Goodwin, P. and Wright, G. (1991). *Decision Analysis for Management Judgment*. John Wiley, Chichester. ISBN 0-471-92833-X.

Clear account of practical multicriterion analysis, and good treatment of uncertainty and probabilities.

'Green Book': see H.M. Treasury (2003).

Green, C.H. and Tunstall, S.M. (1991). The evaluation of river water quality improvements by the contingent valuation method. *Applied Economics* **23**(7), 1135–1146.

Practical advice and commentary.

Green, C.H., Tunstall, S.M. and House, M.A. (1989). Investment appraisal for sewerage schemes: benefit assessment. In *River Basin Management* (Laikari, H. ed.). Pergamon Press, Oxford.

Paper is also available as Publication 148 from the Flood Hazard Research Centre, Middlesex University, UK.

Green, C.H., Tunstall, S.M., Herring, M. and Sawyer, J. (1993). *Customer Preferences and Willingness to Pay for Selected Water and Sewerage Services*. Publication 233. Flood Hazard Research Centre, Middlesex University, UK.

Hanley, N. and Barbier, E.B. (2009). *Pricing Nature: Cost–Benefit Analysis and Environmental Policy*. Edward Elgar, Cheltenham. ISBN 9781-84844-470-6. Sensible and thorough discussion, including commentary on US and UK government guidelines, plus many case studies. Good on sustainability and resilience.

Hanley, N. and Spash, C.L. (1993). *Cost–Benefit Analysis and the Environment*. Edward Elgar, Aldershot. ISBN 1-85278-947-6.

A thorough and fairly academic text on the application of CBA to environmental matters.

Harberger, A.C. and Jenkins, G.P. (2002). *Cost–Benefit Analysis*. Edward Elgar, Aldershot.

An academic volume reprinting 32 papers dating from 1963 to 1999.

Hardy, S.C., Norman, A. and Perry, J.G. (1981). Evaluation of bids for construction contracts using discounted cash flow techniques. *Proceedings of the Institution of Civil Engineers, Part 1* **70**: 91–111.

Hertz, D.B. and Thomas, H. (1983). *Risk Analysis and its Applications.* John Wiley, Chichester. ISBN 0-471-10145-1.

A thorough explanation, with case studies.

Herz, D.B., Subbarao, K., Habib, M. and Raney, L. (1991). *Letting Girls Learn.* World Bank Discussion Paper 133, World Bank, Washington, DC. ISBN 0-8213-1937-X.

A discussion of female education, including the limited use of CBA and a discussion of further benefits, quantified but not valued in economic terms.

HMSO (1991). *Policy Appraisal and the Environment, a Guide for Government Departments.* Her Majesty's Stationery Office, London. (See DCLG, 2009.) ISBN 0-11-752487-5.

H.M. Treasury (2003). *The Green Book; Appraisal and Evaluation in Central Government.* Her Majesty's Treasury, London. ISBN 0-11-560107-4.

The official guide for the British public sector, periodically updated. Available at: http://www.hm-treasury.gov.uk/data_greenbook_index.htm. Places CBA in a wide evaluation context. Relevant commentary is in Hanley and Barbier (2009).

Irvin, G. (1978). *Modern Cost–Benefit Methods.* Macmillan, London. ISBN 0-333-23208-9.

A useful general introduction to CBA from an economist's viewpoint, strong on the economic/social distinction and definition of numéraires.

James, D.E. (1994). *The Application of Economic Techniques in Environmental Impact Assessment.* Kluwer Academic, Dordrecht. ISBN 0-7923-2721-7.

Based on reports prepared for United Nations and Australian agencies for the Asian-Pacific-Australian region. Four chapters on methods, six case studies.

Johnes, G. (1993). *The Economics of Education.* Macmillan, Basingstoke.

A thorough text on education economics. ISBN 0-333-56836-2.

Jones-Lee, M.W. (1989). *The Economics of Safety and Physical Risk.* Basil Blackwell, Oxford. Reprinted, in part, in Layard and Glaister (1994).

Kirkpatrick, C. and Weiss, J. (eds) (1996) *Cost–Benefit Analysis and Project Appraisal in Developing Countries.* Edward Elgar, Cheltenham. ISBN 1-85898-346-0.

A useful collection of papers from a conference in 1995, several of which give accounts of recent thoughts on CBA.

Klümper, S.-A. (1995). *Analysis of Water–supply Projects in Practice.* Conference, University of Bradford, April. Reprinted in Kirkpatrick and Weiss (1996).

Kuiper, E. (1971). *Water Resources Project Economics.* Butterworths, London. ISBN 0-408-70142-0.

A clear explanation by a very articulate engineer of the particular aspects of CBA for water resources projects, but dated and weak on the economic/financial distinction.

Laikari, H. (ed.) (1989). *River Basin Management*. Pergamon Press, Oxford. ISBN 0080-3737-98.

A collection of conference papers, including some on CBA for flood alleviation and wastewater.

Layard, R. and Glaister, S. (eds) (1994). *Cost–Benefit Analysis*. Cambridge University Press, Cambridge. ISBN 0-521-46674-1.

A collection of significant published papers, with a 50-page introduction by the editors explaining CBA from an economist's viewpoint.

Lee, K. and Mills, A. (eds) (1983). *The Economics of Health in Developing Countries*. Oxford University Press, Oxford. ISBN 0-19-261385-5.

Cost-effectiveness analysis, CBA and other economic methods in the health field.

Lind, R.C. (1999). *Analysis for Intergenerational Decisionmaking*. In: Zerbe (2008).

Discusses discounting for long-term projects covering several generations.

Little, I.M.D. and Mirrlees, J.A. (1974). *Project Appraisal and Planning for Developing Countries*. Heinemann Educational, London. ISBN 0-435-84501-2.

Livingstone, I. and Tribe, M. (1995). Projects with long time horizons: their economic appraisal and discount rate. *Project Appraisal* **10**(2), 66–76. See also: *Project Appraisal* 1996, **11**(2), 133–139.

Markandya, A. and Pearce, D. (1994). Natural environments and the social rate of discount. In: *The Economics of Project Appraisal and the Environment* (J. Weiss, ed.). Edward Elgar, Aldershot.

Massam, B.H. (1988). Multi-criteria decision making (MCDM) techniques in planning. *Progress in Planning* **30**(1), 1–84.

A broad summary of multicriterion analysis methods.

Merrett, S. (1997). *Introduction to the Economics of Water Resources – An International Perspective*. UCL Press, Basingstoke. ISBN 1-85728-637-5.

Aimed at engineers, geographers and environmentalists, and assumes no prior knowledge of economics or hydrology. Covers water supply, wastewater, dams and hydropower. Includes case studies.

Mills, A. and Lee, K. (eds) (1993). *Health Economics Research in Developing Countries*. Oxford University Press, Oxford. ISBN 0-19-261620-X.

Fourteen papers on health economics, mainly policy and priorities but with some relevance to CBA.

Morrison, G.C. and Gyldmark, M. (1992). Appraising the use of contingent valuation. *Health Economics* **1**: 233–243.

Also discussion in *Health Economics* 1993, **2**: 357–365.

Murray, C. and Lopez, A. (eds) (1994). *Global Comparative Assessment in the Health Sector: Disease Burden, Expenditures and Intervention Packages*. World Health Organization, Geneva.

NOAA (1993). Report of the NOAA (National Ocean and Atmospheric Administration) panel on contingent valuation. *Federal Register* **58**(10), 4601–4614.

273

Triggered by the *Exxon Valdez* oil spill in Alaska, this reports a thorough review of the contingent valuation method for valuing environmental impacts. Panel led by Kenneth Arrow and Robert Solow and is sometimes referred to under one or both names. Conclusions are summarised in OECD (1995) and Hanley and Spash (1993).

Nutt, P.C. (1989). *Making Tough Decisions – Tactics for Improving Managerial Decision Making*. Jossey-Bass, San Francisco, CA. ISBN 1-55542-138-5.

Decision rules discussed in the context of a private firm.

ODA (1988). *Appraisal of Projects in Developing Countries*, 3rd edn. Overseas Development Administration, London. ISBN 0-11-580256-8.

A practical guide, with case studies. Uses the foreign exchange numéraire.

OECD (1968). *Manual of Industrial Project Analysis*, Vols I and II. Organisation for Economic Cooperation and Development, Paris.

OECD (1995). *The Economic Appraisal of Environmental Projects and Policies, a Practical Guide*. Organisation for Economic Cooperation and Development, Paris. ISBN 92-64-14583-4.

A clear, fairly brief and generally practical guide to valuation methods, including dealing with time and uncertainty (written by J.T. Winpenny).

Parker, D.J., Green, C.H. and Thompson, P.M. (1987). *Urban Flood Protection Benefits – A Project Appraisal Guide*. Gower Technical Press, London.

The *Red Manual*, second in the series from the Flood Hazard Research Centre, Middlesex University. Updates many aspects of the *Blue Manual* (Penning-Rowsell and Chatterton, 1977) and extends the coverage of urban flood alleviation benefits; general principles, plus detailed data for valuing flood damages and losses in the UK.

Pearce, D.W. (1983). *Cost–Benefit Analysis*, 2nd edn. Macmillan, London. ISBN 0-333-35281-5.

Rewritten replacement of popular 1971 book: a basic 110-page text on CBA for economics students.

Pearce, D.W. (1993). *Economic Values and the Natural World*. Earthscan Publications, London. ISBN 1-85383-152-2.

A concise (120-page) discussion of the valuation of environmental impacts.

Pearce, D.W. (1994). Assessing the social rate of return from investment in temperate zone forestry. In: *Cost–Benefit Analysis* (R. Layard and S. Glaister, eds). Cambridge University Press, Cambridge.

A clear account of CBA in the forestry field, covering its special problems of long-delayed benefits and significant environmental aspects.

Pearce, D.W. (1995). Integrating environment into development planning: where have we got to? *Conference on Development Projects: Issues for the 1990s*, University of Bradford, April.

Reviews the practical application of the various ways of valuing environmental impacts, and gives some results.

Pearce, D.W. and Turner, R.K. (1990). *Economics of Natural Resources and the Environment*. Harvester Wheatsheaf, New York. ISBN 0-7450-0225-0.

A 380-page undergraduate textbook on environmental economics and policy, covering CBA alongside wider topics.

Penman, A. (1999). *Financial, Economic and Distributional Analysis – Review Paper Version 2*. Contributing Paper for Thematic Review III.1, as supporting documentation to *Dams and Development*, World Commission on Dams. Earthscan London. ISBN 1-85383-798-9.

Penning-Rowsell, E.C. and Chatterton, J.B. (1977). *The Benefits of Flood Alleviation – A Manual Of Assessment Techniques*. Saxon House, Farnborough. ISBN 0-566-00190-8.

The *Blue Manual*, first in the series from the Flood Hazard Research Centre, Middlesex University. Sets out in detail the methods and procedures for evaluating benefits of flood alleviation, especially regarding damage to property, with data for the UK.

Penning-Rowsell, E.C., Green, C.H., Thompson, P.M., Coker, A.M., Tunstall, S.M., Richards, C. and Parker, D.J. (1992). *The Economics of Coastal Management – A Manual of Benefit Assessment Techniques*. Belhaven Press, London. ISBN 1-85293-161-2.

The *Yellow Manual*, third in the series from the Flood Hazard Research Centre, Middlesex University. Extends the Centre's work on floods and brings in coastal protection and management projects.

Perkins, F.C. (1994). *Practical Cost–Benefit Analysis – Basic Concepts and Applications*. Macmillan Education, NSW. ISBN 0-7329-2783-8.

A thorough textbook with an economics approach relying on supply–demand curves. Comprehensive treatment of financial analysis before starting on economic CBA. Thorough on valuation of labour, environmental aspects, public goods, and distribution adjustments.

Postle, M. (1993). *Development of Environmental Economics for the NRA*. R&D Report 6, National Rivers Authority, Bristol. ISBN 1-873160-47-X.

A brief summary (10 pages) of research activities in the water sector.

Potts, D. (1996). When prices change: consistency in the financial analysis of projects. *Project Appraisal* **11**(1), 27–40.

Potts, D. (2002). *Project Planning and Analysis for Development*. Lynne Rienner, Boulder, CO. ISBN 9781-55587-656-2.

A broad-ranging book on project appraisal, not just by CBA, with emphasis on identifying cost and benefits to particular groups; incentives for commercial firms and for small-scale producers.

Prescott, N. (1993). Cost–effectiveness analysis of chemotherapy regimes of schistosomiasis control. In: *Health Economics Research in Developing Countries* (Mills, A. and Lee, K., eds). Oxford University Press, Oxford.

An example of the complex analysis needed for health interventions, when outcomes depend on behavioural and economic parameters as well as medical linkages.

Price, C. (1993). *Time, Discounting and Value*. Blackwell, Oxford. ISBN 0-631-17986-0.

Price thoroughly examines the technical and ethical details of the rationale for discounting and gives many arguments against it, then discusses other ways that allocation decisions might be made without discounting.

Priemus, H., Flyvbjerg, B. and van Wee, B. (2008). *Decision-Making on Mega-Projects; Cost–Benefit Analysis, Planning and Innovation*. Edward Elgar, Cheltenham.

Thorough discussion of ways to include wide-ranging costs and benefits, especially for transport projects.

Psacharopoulos, G. (1987). *Economics of Education, Research and Studies*. Pergamon Press, Oxford. ISBN 0-08-033379-6.

A collection of papers on economics of education, incorporating some of the World Bank's experience in the field.

Psacharopoulos, G. and Woodall, M. (1985). *Education for Development, an Analysis of Investment Choices*. Oxford University Press, Oxford. ISBN 0-19-520478-6.

Chapter 3 gives a 40-page account of CBA for educational investments, with reference to World Bank experience.

Ray, A. (1984). *Cost–Benefit Analysis – Issues and Methodologies*. Johns Hopkins University Press, Baltimore, OH (for the World Bank). ISBN 0-8018-3069-9.

A fairly academic review of the methodological issues of importance in the early 1980s.

Roberts, J. (1996). Projects with long time horizons: a rejoinder. *Project Appraisal* **11**(2), 133–135.

Rogers, M.G. and Bruen, M.P. (1995). Non-monetary based decision-aid techniques in EIA – an overview. *Proceedings of the Institution of Civil Engineers, Municipal Engineering* **109**: 98–103.

Sassone, P.G. and Schaffer, W.A. (1978). *Cost–Benefit Analysis – A Handbook*. Academic Press, New York. ISBN 0-12-619350-9.

A thorough book using the terminology and mathematics of economics.

Schofield, J.A. (1987). *Cost–Benefit Analysis in Urban and Regional Planning*. Unwin Hyman, London. Revised. paperback edition 1989. ISBN 0-04-445683-2.

A clear text aimed at upper university students of geography and planning, but using economist's language. Good on uncertainty.

Snell, M.J. (1994). Decision aids for planning in the real world. *Journal of the Institution of Water and Environmental Management* **8**(4).

Squire, L. and van der Tak, H.G. (1975). *Economic Analysis of Projects*. Johns Hopkins University Press, Baltimore, OH (for the World Bank). ISBN 0-8018-1818-4.

One of the key publications defining modern methodology in the 1970s, bringing together the UNIDO and OECD approaches.

Stansbury, J., Woldt, W., Bogardi, I. and Bleed, A. (1991). Decision support system for water transfer evaluation. *Water Resources Research* **27**(4), 443–451.

Torrance, G.W. (1986). Measurement of health utilities for economic appraisal, a review. *Journal of Health Economics* **5**: 1–30. Reprinted in: Culyer, A.J. (ed.) (1991). *The Economics of Health*, vols I and II. Edward Elgar, Aldershot.

UNIDO (1972). *Guidelines for Project Evaluation*. United Nations Industrial Development Organisation, Vienna.

van Pelt, M.J.F. (1993). *Ecological Sustainability and Project Appraisal – Case Studies in Developing Countries*. Avebury, Aldershot. ISBN 1-85628-634-7.
A clear account of CBA with application to environmental aspects and the multicriterion context.

van Pelt, M.J.F. (1994). *Financial and Economic Appraisal for Non-economists*. Netherlands Economic Institute, Rotterdam.
A clear explanation of CBA, with sections on government policy instruments also.

Vincke, P. (1992). *Muticriteria Decision-aid*. John Wiley, Chichester. Translation of *L'Aide Multicritère à la Décision*. Editions de l'Université de Bruxelles, Brussels, 1989. ISBN 0-471-93184-5.

Wagstaff, A. (1991). Health care: QALYs and the equity–efficiency tradeoff. *Journal of Health Economics* **10**: 21–41. Reprinted in Layard, R. and Glaister, S. (eds) (1994) *Cost–Benefit Analysis*, Cambridge University Press, Cambridge.

Ward, W.A. and Deren, B.J. (1991). *The Economics of Project Analysis*. World Bank (EDI Technical Manuals), Washington, DC. ISBN 0-8213-1751-2.
Aimed primarily at professional economists, this book starts from market theory and is thorough and precise on numéraires, planning objectives, efficiency and equity. It refers back to earlier World Bank books such as Squire and van de Tak (1975) and Gittinger (1982).

Watson, S.R. and Buede, D.M. (1987). *Decision Synthesis, the Principles and Practice of Decision Analysis*. Cambridge University Press, Cambridge. ISBN 0-521-31078-4.
A thorough text on the principles and psychology of decision-making, including multicriterion decision aids.

Weiss, J. (ed.) (1994a). *The Economics of Project Appraisal and the Environment*. Edward Elgar, Aldershot. ISBN 1-85278-678-7.
Several useful articles.

Weiss, J. (1994b). Double-counting, double not-counting and distribution weights. *Project Appraisal* **9**(4), 283–286.

Whittington, D. and Swarna, V. (1994). *The Economic Benefits of Potable Water Supply Projects to Households in Developing Countries*. Asian Development Bank Staff Paper No. 53. Asian Development Bank, Manila. ISSN 0116-273X.

Willis, K.G. and Corkindale, J.T. (eds) (1995). *Environmental Valuation – New Perspectives*. CAB International, Wallingford. ISBN 0-85198-966-7.
Fifteen papers from two conferences, covering the principles underlying CBA in general and environmental valuation in particular.

Winpenny, J.T. (1991). *Values for the Environment, a Guide to Economic Appraisal*. Her Majesty's Stationery Office, London. ISBN 0-11-580257-6.
A practical and thorough guide, more comprehensive than the author's later book for OECD (1995).

Winpenny, J.T. (1994). *Managing Water as an Economic Resource*. Routledge, London. ISBN 0-415-10378-9.
A thorough and challenging review of policy aspects of the value of water, the section on price elasticity being relevant to CBA. See also OECD (1995), by the same author.

World Bank (1993). *World Development Report 1993 – Investing in Health*. Oxford University Press, Oxford. ISBN 0-19-520890-0.
Emphasis on health economics, especially DALYs.

World Bank (1996). *Handbook on Economic Analysis of Investment Operations*. Unpublished draft, May.
The purpose of this handbook, and of a modified version dated 1998, was to provide Bank staff and others with practical but solidly grounded analytical tools, and to make project evaluations transparent. Aimed at economists, it systematically covered the identification and valuation of costs and benefits. In 2001 it was superseded by the published version, Belli *et al.* (2001).

Yaffey, M. (1992). *Financial Analysis for Development*. Routledge, London. ISBN 0-415-08095-9.
A clear explanation of the financial analysis of projects and enterprises, including financial CBA, especially for developing countries.

Yaffey, M. (1994). Old and new thinking on key appraisal indicators. *Project Appraisal* **9**(2), 119–126.
A review of recent thinking, with special emphasis on the net benefit to capital ratio.

Yaffey, M. and Tribe, M.A. (1992). *Project Rehabilitation in Adverse Environments*. Avebury, Aldershot. ISBN 1-85628-341-0.
A useful introduction to the particular problems of rehabilitation projects.

Zeleny, M. (1982). *Multiple Criteria Decision Making*. McGraw-Hill, New York.
A comprehensive (550-page) book on the subject, with broad thinking and many practical insights. Multiple ISBN 0-07-072795-3.

Zerbe, R.O. (2004). *Should Moral Sentiments be Incorporated into Benefit–Cost Analysis? An Example of Long-term Discounting*. In: Zerbe (2008).

Zerbe, R.O. (ed.) (2008) *Benefit–Cost Analysis*. Edward Elgar, Cheltenham. ISBN 978-1847-20-964-1.
Two fat volumes reprinting numerous papers, mostly of an academic nature and emanating from the USA.

Zerbe, R.O. and Bellas, A.S. (2006). *A Primer for Benefit–Cost Analysis*. Edward Elgar, Cheltenham. ISBN 1-84376-897-6.
A scholarly exposition of CBA in an exclusively US context. Some confusion between 'interest rate' and 'discount rate'.

Zerbe, R.O. and Dively, D.D. (1994). *Benefit–Cost Analysis in Theory and Practice*. HarperCollins, New York. ISBN 0-673-18066-2.
A thorough economics text referring especially to public policy and expenditure in the USA.

Journals

The following are some of the relevant journals, which occasionally carry papers or book reviews related to CBA.

General

Impact assessment and project appraisal. Beech Tree Publications, Guildford. ISSN 1461-5517 (before 1998 it was *Project Appraisal*, ISSN 0268-8867).
Applied Economics, Routledge, UK. ISSN 0003-6846 (online 1466-4283).
Finance and Development, International Monetary Fund, USA. ISSN 0145-1947 (formerly 0015-1947).
World Development Report, annual, World Bank; each year's issue has a specific theme. ISSN 0163-5085.
Asian Development Bank, Staff Papers appear from time to time (e.g. Whittington and Swarna (1994) listed above). ISSN 0116-273X.

Engineering

Proceedings of the Institution of Civil Engineers (UK).
Civil Engineering (ISSN 0965-089X), *Municipal Engineering* (ISSN 0965-0903), *Transport* (ISSN 0965-092X), and other specialist sections, Thomas Telford Publishing, UK.

Health

Journal of Health Economics, Wiley, UK. ISSN 1057-9230 (online 1099-1050).
Health Policy and Planning, Oxford University Press and London School of Hygiene and Tropical Medicine, UK. ISSN 0268-1080 (online 1460-2237).
International Journal of Health Planning & Management. ISSN 0749-6753 (online 1099-1751).

Water resources

Water and Environment Journal, Chartered Institution of Water and Environmental Management, UK. ISSN 1747-6585 (formerly 0951-7259).
Water Resources Research, American Geophysical Union, USA. ISSN 0043-1397).

Planning
Progress in Planning, Elsevier. ISSN 0305-9006.

Multicriterion analysis
Journal of Multi-Criteria Decision Analysis, Wiley. ISSN 1057-9214 (online 1099-1360).

Glossary/index

'When I use a word', Humpty Dumpty said, 'it means just what I choose it to mean, neither more nor less'.

This combined index and glossary lists the main key words that may help readers to find particular topics in the book, and gives a brief definition or explanation of some of them. Where there is no explanation or a reader needs a fuller one (some concepts can only be explained in their context), reference should be made to the indicated page or section. If there are several references, the main one may be in bold type. References to other entries are usually in capitals.

Acceptable risk analysis, 158

Accounting prices, 27. Another name, not generally used in this book, for ECONOMIC PRICES.

Accounting rate of interest. A term, not generally used in this book, for the DISCOUNT RATE.

ADB, 25–6. ASIAN DEVELOPMENT BANK.

Agriculture, 82–90, 131 Fig. 5.5

Aircraft noise, 186

Analysis period, 52–4 sect. 3.4, 135, 153, 239. The sequence of years over which a project is analysed.

Analyst, role of, 7–8

Analytic hierarchy process, 228

Asian Development Bank (ADB), 25–6. One of the leading financing agencies for development projects.

Average incremental cost, 95

Average unit cost, 33 Box 2.5. The total cost of producing something divided by the number of units produced, cf. MARGINAL UNIT COST.

Avertive behaviour, 185–6

Base case, 61. The set of BEST-ESTIMATES of parameters defining a CBA, as a starting point for SENSITIVITY TESTS.

B/C ratio, 46, sect. 3.3. BENEFIT/COST RATIO.

Benefit, 1, **13–16** sect. 2.1 Any impact or consequence of a course of action which is desirable or to be maximised.

Benefit–cost analysis. An alternative name for COST–BENEFIT ANALYSIS.

Benefit/cost ratio (B/C ratio), 46 sect. 3.3. The ratio of the PRESENT VALUE (PV) of all benefits to the PV of all costs, for a project or alternative subjected to CBA; one of the INDICATORS used in CBA.

Benefit transfer, 90, 106, **189–90** sect. C.8 Use of benefit estimates calculated in one place for an analysis in another place.

Bequest value, 175 Box C1

Best-estimate, 20–2 Box 2.3. An analyst's estimate of the expected value of a future quantity or number, not biased by any desire to avoid an underestimate.

Bimodal, 144, Box A2

Biological habitats, 188

Black box, 113, 213. A device that receives inputs and produces outputs but does not clearly reveal the processes that link them.

Border price, 167, 245. The price of an import or export at a nation's border, which is the nation's OPPORTUNITY COST for that item and is therefore the relevant starting point for valuing imports, import substitutes and exports for an ECONOMIC ANALYSIS.

Border rupees, 165

By-products, 83

Capital rationing, 70–1, **203–4** sect. E.4 One of the roles of the DISCOUNT RATE.

Cash flow, 11, 136. The list of NET BENEFITS in successive years.

Cash margin, 88

CBA. Cost–benefit analysis.

Central measure, 143 Box A2. A number indicating the central value or tendency of a PROBABILITY DISTRIBUTION.

Checklist, 134 ch. 6

Climate change, 149, 152, 155

Coastal protection, 98–110 sect. 4.6

COBA, 112–14. A standardised package for CBA of road schemes in UK.

Coefficient of variation, 143 Box A2. A DISPERSION MEASURE for PROBABILITY DISTRIBUTIONS; the STANDARD DEVIATION divided by the MEAN.

Common property good, 174 Box C1

Common unit of measurement (numéraire), 2, 24–6 sect. 2.3, 163. A measuring unit, usually a currency, used to quantify dissimilar things for numerical comparison.

Compensation, 180

Components (separable project components), 134

Composite programming, 231 Note 9

Compromise, 226. See MULTICRITERION ANALYSIS.

Concordance (index), 227–8

Consensus maximisation, 231 Notes 10–11

Constant prices, 54–5, 68. The unit used to express sums in

constant prices is a currency at a defined moment in time, so that the price levels of that moment apply for costs and benefits throughout the ANALYSIS PERIOD, and INFLATION is not relevant to the analysis. Amounts expressed in constant prices are sometimes said to be in REAL TERMS.

Constant-value terms, 68. Another word for REAL TERMS or IN CONSTANT PRICES.

Consumer surplus, 175, 193 Note 17

Consumption rate of interest, 202, 210 Note 14

Contingency allowances contingencies, 20–2, 153

Contingent ranking, 182

Contingent valuation, 182–4, 193 Note 12

Continuous distribution, 144, Box A2

Conversion factors, 39 Note 13, 165, 171 Notes 8–9, 177

Convertible currency, 166

Correlation, 154, 156

Cost, 1, **13–16** sect. 2.1. Any impact or consequence of a course of action which is undesirable or to be minimised.

Cost–benefit analysis (CBA), 1, 4 Table 1.1. A tool for guiding decisions between courses of action where the COSTS and BENEFITS are expressed in a COMMON UNIT OF MEASUREMENT.

Cost-effectiveness analysis, 3, 4 Table 1.1, 91, 94. A tool for guiding decisions between

courses of action where the COSTS and BENEFITS are not expressed in a COMMON UNIT OF MEASUREMENT, but the costs per unit of benefit are compared.

Cost estimates, 21, 31

Cost recovery, 89, 246

Cost-of-risk, 100, 142. The DAMAGE-LOSS caused by an event (such as a flood or earthquake of a particular size) multiplied by its probability of occurring.

Cost–utility analysis, 92. A device for quantifying costs and benefits of health projects or health-related decisions.

Critical load, 158

Crop budget, 83, 84 Table 4.1, 86 Table 4.3, 246 Table W3, 247 Table W5

Current prices, 55. Sums expressed in current prices are each at the price level of the moment of occurrence; in the presence of INFLATION the current prices of different years are not the same unit and therefore cannot be usefully added together for analysis: see CONSTANT PRICES.

DALYs, 92 Box 4.2. DISABILITY ADJUSTED LIFE-YEARS.

Damage-loss, 99, 101–2. A term used in this book for the sum of all costs resulting from a flood, storm, earthquake or suchlike event; includes the cost of repairing damage and all quantified direct and indirect costs resulting from the event.

Damage and response functions. A kind of DOSE–RESPONSE FUNCTION.

Decision-makers, 7–8 sect. 1.6, 121, 149, 193 Note 24, 206, 212–13. The people responsible for making, and usually also for defending and implementing, a decision.

Decisions, types of, 3

Decommissioning cost, 53, 75, 239–0

Defensive expenditure, 185–6

Demand forecasts, 80, 117, 125

Depletion premium, 35. An element of economic cost representing the contribution to depletion of a finite resource.

Depreciation, **22** sect. 2.2.6, 24 Table 2.1, 55. Sums of money set aside each year to provide for the later replacement of things which are wearing out; an accounting device not representing the real use or production of anything, and therefore not counted in CBA.

Design-type decision, 3. See EITHER/OR DECISION.

Diminishing marginal utility of consumption, 36, 201

Direct use value, 174 Box C1

Disability-adjusted life-years (DALYs), 92 Box 4.2. A device for quantifying costs and benefits of health projects or health-related decisions.

Discontinuities, 158

Discordance index, 227

Discount rate
choice of, 68–75, 200–9 **App. E**
definition, 42, 209–10 Note 14

Discounted total volume (of water), 46

Discounting, 11, **42–6** sect. 3.2, 257–66. Adjusting the numerical valuation of costs, benefits or other resources to reflect the time at which they occur in a project's lifespan; see also TIME PREFERENCE.

Discrete distribution, 144, Box A2

Dispersion measure, 143 Box A2. A number indicating the degree of spread in a PROBABILITY DISTRIBUTION.

Displaced production, 15. See SUBSTITUTION.

Distance-from-ideal approaches, 225–6

Distribution of costs and benefits between different people, 36, 195 App. D

Do-nothing scenario, 14. A particular kind of WITHOUT-PROJECT SITUATION.

Domestic price level, 167, **sects B.3 to B.5**

Domestic pricing numéraire, 25, 26, **164–5** sect. B.3, 171 Notes 10–11. One of the two common definitions of the COMMON UNIT OF MEASUREMENT for ECONOMIC ANALYSIS: costs and benefits are valued in terms of their impact on society's consumption,

measured as people's
WILLINGNESS-TO-PAY.

Dominance, 217–18

Dose–response functions, 90, 111, **173**. Technical linkages between a project output or impact and some other good which is easier to put a value on.

Double-counting, 36–7

Earnings differentials, 187

Economic analysis, 5, 18. A CBA done on behalf of or from the point of view of a defined group of people, usually a nation. The COMMON UNIT OF MEASUREMENT is related to economic efficiency, so that money paid or received is not necessarily an adequate measure of costs and benefits; ECONOMIC PRICES are used to adjust for market distortions and/or to reflect economic efficiency (cf. FINANCIAL ANALYSIS, SOCIAL CBA, ECONOMIC APPRAISAL)

Economic appraisal, 6. Assessment of a project's economic merit and impact; often implies something wider than ECONOMIC ANALYSIS, including also FINANCIAL ANALYSIS from the point of view of key agents so as to investigate their MOTIVATION for taking part in the project.

Economic price, 27, **172**, 243–5. A measure of the value of a good or service to a defined

group of people (usually a nation), in terms of a defined COMMON UNIT OF MEASUREMENT. Often arrived at from a FINANCIAL PRICE by SHADOW PRICING.

Education, **96–8** sect. 4.5, 187

Effects method, 26, 232 App. G, 233 Notes 1–2. A special kind of CBA, seldom used outside French-speaking countries, which uses coefficients that describe linkages between parts of an economy and traces changes in value-added caused by a project.

Efficiency pricing, 5, 29. Prices adjusted to correspond to a hypothetical perfect market.

EIRR, 48. Economic INTERNAL RATE OF RETURN, one of the INDICATORS used in CBA.

Either/or decision, 3. Decision between alternative courses of action.

Eliciting, 155

Energy, 115–17 sect. 4.8

Environment, 98–110 sect. 4.6

Equalising discount rate, 61. A DISCOUNT RATE which causes the NPVs of two alternatives to be equal.

Equity between generations, 68, 205–6

Escalation, 55, 135. An increase in the value or price of something, even in REAL TERMS (CONSTANT PRICES); its value in CURRENT PRICES increases by more than the rate of INFLATION.

Existence value, 175 Box C1.

Expected value, 143 Box A2. Arithmetic mean of a PROBABILITY DISTRIBUTION.

Externality, 17 Box 2.1. (An ambiguous term, see Box 2.1) a Sort of NON-MARKET COST OR BENEFIT.

Exxon Valdez oil spill, 193 Note 15

Family labour, 83, 86, 178

Farm budget, 84, 85 Table 4.2, 247 Table W4, 248 Table W6

FCERM-AG and FCDPAG, British flood appraisal guidelines, 104, **119** Note 23

Feasibility study, 9

Feasible, 67

Financial CBA (or financial analysis), 4, 127, 128 Fig. 5.2. A CBA from the point of view of an individual or entity, accepting actual money spent or received as the sole unit of measurement (cf. ECONOMIC ANALYSIS). (Financial project analysis, in a wider sense, is outside the scope of this book: it is covered by Yaffey (1992).)

Financial price, 27. The money price actually paid for a good or service.

Finite resources, 35

FIRR, 48. Financial INTERNAL RATE OF RETURN.

Fixed costs, 33, Box 2.5

Fixed trip matrix, 113

Flexibility, 152, 159. This is a valuable characteristic, making a project more robust and less vulnerable to uncertainty.

Flood control and alleviation, 98–110 sect. 4.6

Foreign exchange numéraire, 25, **165–6**, 170 Note 6. One of the two common definitions of the COMMON UNIT OF MEASUREMENT for ECONOMIC ANALYSIS: costs and benefits are valued in terms of free foreign exchange in the hands of the government, though usually expressed in local currency using an official exchange rate.

Forestry, 90, 206

Free market, 28, 176–7. A market in which the forces of supply and demand are allowed to fix prices without interference or distortion.

Free rider problem, 183

Fuel wood, 181–2

Future value (FV), 41. The value of some amount of resources before adjustment for TIME PREFERENCE by DISCOUNTING.

FV, 41. FUTURE VALUE.

Gaussian distribution, 144, Box A2

Goal achievement matrix, 228

Government failure, 177

Government guidelines, British, 72, 160 Note 1

Green Book, The, 25, 272. The official guide to CBA in the British public sector, published by H.M. Treasury and periodically updated (H.M. Treasury, 2003).

Gross income, 88

Gross margin, 83

Gross/net ratio, 76 Note 12
Gross output, 88
Gross return, 83, 88

HDM-III, 115. A standardised
model for estimating vehicle
operating costs in relation to
road design, maintenance and
surfaces.
Health, **90–6** sect. 4.3 and 4.4,
187
Healthy years of life gained
(HYLGs), 92 Box 4.2. A
device for quantifying costs
and benefits of health projects
or health-related decisions.
Hedonic pricing, 186
Hierarchies, 218, 230 Note 3
Histogram, 143 Box A2, 145 Fig.
A1. Graph showing the shape
of a PROBABILITY DISTRIBUTION.
Household production function,
186
Human capital, 97, 187
HYLGs, 92 Box 4.2. See Healthy
years of life gained.

Income weighting, 196–8 sect.
D.2, 197 Box D1
Incremental costs/benefits,
10–11, 88, 248 Table W7, 249
Table W8. In a CBA,
incremental normally means
the difference between the
costs/benefits in the WITH-
PROJECT SITUATION, and those in
the WITHOUT-PROJECT SITUATION.
Incremental net returns, 88
Indicators, 46–52 sect. 3.3,
57–60 sect. 3.6. Numbers used
to express the relative or

absolute merit of a project or
alternative; the numerical
output of a CBA calculation:
see NPV, B/C RATIO, IRR, N/K.
choice of (summary), 60
Indifference, 40. Absence of any
preference between two states.
Indirect use value, 174 Box C1
Inflation, 54–7 sect. 3.5. The
decrease in the value of a
currency with time; see
CONSTANT PRICES.
Information, value of, 152, 161
Note 10
Insurance, 151
Intangible, 17. A vague term
usually meaning a project
impact which, for whatever
reason, is not included in a
CBA; see EXTERNALITY.
Interest (payments) on a loan,
22–3, 24 Table 2.1, 55–6
Internal rate of return (IRR), 48.
The discount rate which
causes the NPV to be zero and
the B/C RATIO to be 1.0; one of
the INDICATORS USED in CBA.
Internalise, 18. Bring an effect
that was an EXTERNALITY into
an analysis, so that it ceases to
be one.
Internally optimised, 58, 60
Investment-type decision, 3. See
YES/NO DECISION.
IRR, 48. INTERNAL RATE OF
RETURN.
Irreversibility, 158

Labour, 178, 192 Note 7
Land, 179–80
Landscape, 106–7

Latin Hypercube, 157. A sampling method for RISK ANALYSIS.

Least-bound approach. See LOWER BOUND.

Least-cost analysis, 3, 4 Table 1.1, 58, 116. An analysis between TECHNICALLY MUTUALLY EXCLUSIVE alternatives which all give the same benefits, so that only costs need to be compared.

Less developed economies, 80

Lexicographic ordering methods, 225

Life, valuation of (statistical life), 187, 189

Limited factor analysis, 51

Limited resource, 51

LMST system, 170–1 Note 7

Loans and repayments, 22–3, 24 Table 2.1, 55–6

Locus of switching values, 65

Log–normal probability distribution, 144 Box A2

Lower bound (least-bound), 23–4, **107–10** sect. 4.6.4, 153, 154. To compare a project against a defined threshold it may be sufficient to find the lowest plausible value of a parameter rather than its expected value.

Margin, 88

Marginal unit cost (or marginal cost), 32, 33 Box 2.5. The change in total production cost when one more unit of something is produced, cf. AVERAGE UNIT COST.

Market distortions, 28

Market failure, 176–7

MAUT, 231 Note 11. MULTI-ATTRIBUTE UTILITY THEORY.

Maxi-max rule, 161 Note 8

Maxi-min rule, 152

Mean, 143–4 Box A2. The average of several numbers; the EXPECTED VALUE of a statistical PROBABILITY DISTRIBUTION.

Median, 143–4 Box A2. A CENTRAL MEASURE for PROBABILITY DISTRIBUTIONS; the value which is or is expected to be exceeded half the time or by half the parameter values.

Merit good, 174 Box C1

Migrant labour, 178

Minimax-regret rule, 152

Minimum wage, 178

Miscellaneous items, 21. Items in a cost estimate which refer to small parts of a project which have not yet been detailed at a particular stage of the design process, but are definitely expected to be incurred; sometimes confused with PHYSICAL CONTINGENCIES.

Mitigating measures, 14

Mode, 143–4 Box A2. The most common or likely value or group of values in a PROBABILITY DISTRIBUTION.

Monopoly, 176

Monte-Carlo, 157, 162 Notes 18–19. A sampling method for RISK ANALYSIS.

Morbidity, 91. Incidence of disease, cf. MORTALITY.

Mortality, 91. Incidence of death, cf. MORBIDITY.

Motivation, 79, 84, 88

Multi-attribute utility theory (MAUT), 231 Note 11

Multicriterion analysis, 212 App. F

Multi-objective display matrix, 199 Note 7

Multiplier effects, 36

Multipurpose projects, 79–80

Mutually exclusive, 3, 58, 59 Box 3.5. See TECHNICALLY MUTUALLY EXCLUSIVE.

Myopia, 201

Negative benefits, 88

Negative costs, 52, 66. It is sometimes convenient to treat some classes of benefits as negative costs.

Net benefit, 11, 88. The INCREMENTAL BENEFITS minus the INCREMENTAL COSTS, usually for a particular year in a project's lifetime; see CASH FLOW.

Net benefit/investment ratio (N/K), 51. The PV of NET BENEFITS divided by investment costs; one of the INDICATORS used in CBA.

Net margin, 88

Net present value (NPV), 46, 125–6. The sum of PRESENT VALUES (PV) of all the NET BENEFITS of a project or alternative subjected to CBA, also equal to the PV of all benefits minus the PV of all costs; one of the INDICATORS used in CBA.

Net return, 83, 88

N/K, 51. See NET BENEFIT/INVESTMENT RATIO.

Nominal interest rate, 70 Box 3.6. The interest rate when calculations are done in CURRENT PRICES: it is a product of the REAL INTEREST RATE and the rate of INFLATION.

Nominal prices. Another name for CURRENT PRICES.

Non-excludability, 174 Box C1

Non-market costs/benefits, 17–18 sect. 2.2.3, 24 Table 2.1, 34–5 sect. 2.6.3. Costs or benefits which are not bought and sold and therefore have no market prices.

Non-rival consumption, 174 Box C1

Non-traded or non-tradeable goods and services, 25, 164 sect. B.2, 167 Table B1, 192 Note 10. Goods and services which are not imported or exported, or which by their nature are never likely to be.

Non-use value, 175 Box C1

NPV, 46, 125–6. NET PRESENT VALUE.

Normal distribution, 144, Box A2

Numéraire, 135, 163 App. B. See COMMON UNIT OF MEASUREMENT.

OECD, 165, 170–1 Note 7, 274

OER, 165, 242. OFFICIAL EXCHANGE RATE.

Official exchange rate (OER), 165, 242

Opportunity cost, 29 Box 2.4, 178–82 sect. C.3, 209 Note 8.

A measure of the value of something in terms of what we have to forgo or give up in order to obtain it.

Opportunity cost of capital, 69, **202–3** sect. E.3, 209 Note 8 and 11. The real interest rate (rate of return) on the marginal investment; often used as a guide to the appropriate DISCOUNT RATE, on the assumption that it gives the return forgone in the next best investment.

Optimally packaged, 60

Option value, 175 Box C1

Outranking, 227–8

Parity prices, 181

Passive use value, 175 Box C1

Physical contingencies, 20 Box 2.3, 21–2, 24 Table 2.1, 76 Note 7. An addition to the BEST-ESTIMATE of a future number, in order to produce a consciously biased estimate with a relatively low chance of being exceeded. Appropriate for budgeting but not for inclusion in a CBA (though it often is).

Planned balance sheet, 228

Policy instruments, 188

Precautionary principle, 158

Precision, level of, 10

Prefeasibility study, 9

Preferences, 182–8 sect. C.4 and C.5, 191 Note 1

Present value (PV), 41, 75–6 Note 5. The equivalent value of some amount of resources

after adjustment for TIME PREFERENCE by DISCOUNTING.

Presentation, 120 ch. 5, 239, 241 App. I

Price. The value of one unit of something, whether in a financial, economic or social sense.

Price contingencies, 21, 55, 56 Box 3.4, 76 Note 7. An allowance added to a cost estimate to allow for future inflation; when adding together costs incurred in different years, it is the difference between the sum of costs in CONSTANT PRICES and those in CURRENT PRICES.

Pricing, 5. See FINANCIAL ANALYSIS, ECONOMIC ANALYSIS, SOCIAL ANALYSIS, SHADOW PRICING.

Private good, 174–5 Box C1

Probability distribution, 143–4 Box A2, 145 Fig. A1. Graph or function showing the probability of a certain parameter having values within certain ranges; see EXPECTED VALUE, MEAN, MEDIAN, MODE, RETURN PERIOD.

Project appraisal, 8. See ECONOMIC APPRAISAL.

Proxy markets, 184–8 sect. C.5

Public good, 174–5 Box C1

Public health, 94–6 sect. 4.4

PV, 41, 75–6 Note 5. PRESENT VALUE.

QALYs, 91, 92 Box 4.2, 118 Note 11. QUALITY ADJUSTED LIFE-YEARS.

Quality adjusted life years (QALYs), 91, 92 Box 4.2, 118 Note 11. A device for quantifying costs and benefits of health projects or health-related decisions.

Rank-arithmetic approaches, 226
Ranking, 58, **218–20**. Sorting a number of alternatives or projects into decreasing order of merit.
Rate of growth of per capita consumption, 201
Real interest rate, 70 Box 3.6. The interest rate when calculations are done in CONSTANT PRICES. See NOMINAL INTEREST RATE.
Real terms, 54–5. See CONSTANT PRICES.
Recreation, 98–110 sect. 4.6
Regional planning, 117
Rehabilitation, 9. Improvement of an existing project or infrastructure, usually made necessary by one or more of: decay, wear and tear, inadequate maintenance, changing needs.
Repeated value (RV), 44–5. One of a series of equal FUTURE VALUES.
Replacement costs, 22, **54**, 187–8, 250 Table W9. The costs of replacing or refurbishing those parts of a project which wear out earlier than the project as a whole, especially mechanical and electrical equipment within a civil engineering project.
Residual value, 53. The remaining value of infrastructure at the end of the ANALYSIS PERIOD; usually ignored in CBA calculations provided that the period is long enough to make the PRESENT VALUE of any residual value negligible.
Return to investment, 60. The interest rate earned by a financial investment; related to INTERNAL RATE OF RETURN but not identical except in some kinds of FINANCIAL ANALYSIS.
Return period, 146. For an event such as a flood, earthquake or high wind that is subject to a PROBABILITY DISTRIBUTION, the return period in years is the reciprocal of the probability of exceedance in any one year: for example, a '20-year flood' is one with a return period of 20 years and an exceedance probability in any one year of 005.
Risk, 8, 66, 141–62 App. A, 142 Box A1
Risk analysis, 66, **154–8**, 205
Risk-averse, 150 Figure A3, 160–1 Note 5
Risk–benefit analysis, 158
Risk-neutral, 149, 150 Figure A3
Risk pooling and spreading, 150–1
Risk premium, 71, 153, **204**, 209 Note 13
RV, 44–5. REPEATED VALUE.

Safe minimum standard, 158
Safety, 187
Sanitation, 94–6 sect. 4.4
Scarcity value, 28. Value
    attached to something because
    it is not in plentiful supply; see
    also OPPORTUNITY COST.
Scenario planning, 152
Scoring, 214–17 sect. F.2.2
Screening, 218–20
Sensitivity indicator/index, 62,
    251. In a SENSITIVITY TEST, the
    percentage change in an
    INDICATOR corresponding to a
    1% change in an input
    parameter.
Sensitivity test, 8, 11, 61–6 sect.
    3.7, 129, 131–3 ch. 5, 136–7,
    153–4 sect. A.3.2, 161 Note
    14, 251–3 sect. 1.8. An
    analysis of what would happen
    to the results of a calculation if
    one of its input parameters
    were changed while all the
    others remained the same.
Separable costs – remaining
    benefits method, 80, 81 Box
    4.1. A convention for
    allocating the costs of a
    multipurpose project between
    the purposes.
SER, 165, 170 Note 6, 242.
    SHADOW EXCHANGE RATE.
Shadow exchange rate (SER),
    165, 170 Note 6, 242
Shadow price, 27. Another name,
    not generally used in this book,
    for ECONOMIC PRICE.
Shadow price factor (SPF), 27,
    31–2. Factor sometimes used
    in shadow pricing: a FINANCIAL

PRICE multiplied by an
    appropriate SPF gives an
    ECONOMIC PRICE.
Shadow pricing, 26–31 sect. 2.4
    and 2.5, 188 App. C.
    Estimation of ECONOMIC PRICES,
    usually by adjusting FINANCIAL
    PRICES.
Shadow projects, 106, 187–8
Simulation, 154–8 sect. A.3.3
Size optimisation, 128–9
Social, 30. In many CBA
    contexts 'social' means
    collective as opposed to
    private; as in SOCIAL
    OPPORTUNITY COSTS, SOCIAL TIME
    PREFERENCE RATE. See also
    SOCIAL CBA.
Social CBA, 5–6. A CBA in
    which special VALUATION
    METHODS are used to reflect
    ethical objectives that a purely
    efficiency-oriented ECONOMIC
    ANALYSIS would not reflect
    (often, however, merged with
    the term ECONOMIC ANALYSIS).
Social opportunity cost, 69, 203,
    209 Note 7. Another name for
    the OPPORTUNITY COST OF
    CAPITAL.
Social time preference rate, 69,
    202, 209 Note 7, 210–11
    Note 19
SPF, 27, 31–2. SHADOW PRICE
    FACTOR.
Spreadsheet, 49, 50–1 Box 3.3.
    A self-calculating table set up
    on a computer.
Standard conversion factor,
    165–6
Standard deviation, 143 Box A2.

A popular DISPERSION MEASURE for PROBABILITY DISTRIBUTION.

Subsidies, 16–17. Payments from a government to people or entities within a nation; being TRANSFER PAYMENTS they are not counted in a normal ECONOMIC ANALYSIS.

Substitution, 15. If a project's product substitutes for or displaces a product that was formerly made somewhere else, the effect on the former producer may need to be taken into account in the project's CBA (depending on the former producer's location and on the type of analysis).

Sunk costs, 19–20 Box 2.2, 38 Note 6. Costs already spent or committed before a decision is taken, so that they do not depend on the decision and should not be counted in a CBA that aims to guide it.

Surrogate goods, 186

Sustainable, 67, 106–7

Switching value, 63–5 Figure 3.1, 137, 154, 238. In a SENSITIVITY TEST, the value of a parameter which, if substituted for the parameter's BASE CASE value, will bring the INDICATORS to their critical or threshold values.

Tariff, 95

Taxes, 16–17 Table 2.1. Payments to a government, which as TRANSFER PAYMENTS

are not counted in a normal ECONOMIC ANALYSIS.

Technically mutually exclusive, 3, 59 Box 3.5. Describes alternatives of which only one can be chosen, for technical reasons rather than because of shortage of resources.

Test discount rate, 57. A DISCOUNT RATE that is used for determining INDICATORS that depend on the discount rate (NPV, B/C RATIOS of all sorts), or is fixed as a threshold or reference value against which IRRs are compared.

Time preference, **40–2** sect. 3.1, **68–9, 201–2** sect. E.2. The fact that people prefer to experience good things sooner rather than later; the basis for DISCOUNTING.

Timing of project development, 125–7

Toll good, 174 Box C1

Trade-off, 223. See also MULTICRITERION ANALYSIS.

Traded or tradeable goods and services, 25, 164, 167 Table B1, 180, 192 Note 10. Goods and services which are, or could be, imported or exported.

Transfer payments, 16–17, 24 Table 2.1. Financial payments between parties within a nation, which therefore are not counted in an ECONOMIC ANALYSIS on behalf of that nation, but are counted in a FINANCIAL ANALYSIS on behalf of either of the parties. The main examples are TAXES (including customs duties) and SUBSIDIES.

Transformation function, 214

Transparency, 7–8 sect. 1.6, 213. Desirable feature of an analysis and its reporting, whereby all assumptions, criteria and methods are made explicit and clear to decision-makers and affected parties.

Transport, 111–15 sect. 4.7

Travel cost method, 109, 185

Treasury, British, 60

Triangular, 144, Box A2

Uncertainty, 8, 66, 141–62 App. A, 142 Box A1

Uneconomic, 67

Unemployment, 30. In places with high unemployment the OPPORTUNITY COST of some sorts of labour is low; SHADOW PRICING can then be used to bias decisions so as to help relieve unemployment.

UNIDO, 38 Note 11, 164, 276

Unit cost, 33 Box 2.5. Cost per physical unit: see AVERAGE UNIT COST and MARGINAL UNIT COST.

Unit reference value, 95

Unit value. See PRICE.

Unquantifiables, 18. Project effects that are omitted from a CBA because they are considered difficult or impossible to put a value on; see also EXTERNALITY.

Urban planning, 117

Use value, 174–5 Box C1

Utility function/curve, 151

Valuation methods, 3, 188–9 sect. C.7. Ways of quantifying costs and benefits in terms of a COMMON UNIT OF MEASUREMENT.

Value-added, 232

Value function, 214

Value judgments, 7, 12 Note 5

Value tree, 219 Figure F2

Variable costs, 33, Box 2.5

Variable trip matrix, 113

Viable, 67, 79

Voluntary inputs, 18. Unpaid work or donations to a project motivated by ethical considerations.

Water supply, 94–6 sect. 4.4

Weighted average, 217 sect. F.2.3

Weighted cost-effectiveness analysis, 3. A name for some kinds of MULTICRITERION ANALYSIS.

Willingness to accept compensation (WTA), 182–3. See WILLINGNESS-TO-PAY.

Willingness-to-pay (WTP), 12 Note 5, 28, 34, 91, 93, 95–6, 109, 114, 115–16, 164–5, 172, **175, 182–3**. One of the main bases for valuation in CBA; usually expressed by market prices but sometimes estimated by other means.

With-project situation, 10, 13–16. The assumed situation if a project is implemented; together with the WITHOUT-PROJECT SITUATION this defines the project and enables the INCREMENTAL COSTS AND

BENEFITS to be unambiguously identified.

Without-project situation, 10, 13–16, 108. The assumed situation if a project is not implemented; it is not normally identical to a continuation of the present situation; see WITH-PROJECT SITUATION.

World Bank, 12 Note 9, 38 Note 7, 277–8. One of the leading financing agencies for development projects.

World price level, 25, sects B.3 to B.5. A numéraire based on a convertible foreign currency and on BORDER PRICES; also called FOREIGN EXCHANGE NUMERAIRE.

WTA (Willingness to accept compensation), 182–3. See WILLINGNESS-TO-PAY.

WTP. See WILLINGNESS-TO-PAY.

Years of potential life gained (YLGs), 92 Box 4.2. A device for quantifying costs and benefits of health projects or health-related decisions.

Yes/no decision, 3. Decision whether or not to undertake a defined course of action, e.g. to implement a certain project.

YLGs, 92 Box 4.2. YEARS OF POTENTIAL LIFE GAINED

'That's a great deal to make one word mean', said Alice . . .